T0275954

SpringerBriefs in Immunology

More information about this series at http://www.springer.com/series/10916

David Escors · James E. Talmadge
Karine Breckpot · Jo A. Van Ginderachter
Grazyna Kochan

Myeloid-Derived Suppressor Cells and Cancer

 Springer

David Escors
Navarrabiomed-Biomedical Research
  Centre, Fundación Miguel Servet
IdiSNA
Pamplona, Navarra
Spain

Jo A. Van Ginderachter
VIB Lab Myeloid Cell Immunology,
  Building E
Vrije Universiteit Brussel
Brussels
Belgium

James E. Talmadge
Department of Pathology and Microbiology
University of Nebraska Medical Center
Omaha, NE
USA

Grazyna Kochan
Navarrabiomed-Biomedical Research
  Centre, Fundación Miguel Servet
IdiSNA
Pamplona, Navarra
Spain

Karine Breckpot
Faculty of Medicine and Pharmacy
Vrije Universiteit Brussel
Brussel
Belgium

ISSN 2194-2773   ISSN 2194-2781 (electronic)
SpringerBriefs in Immunology
ISBN 978-3-319-26819-4   ISBN 978-3-319-26821-7 (eBook)
DOI 10.1007/978-3-319-26821-7

Library of Congress Control Number: 2016934670

Printed on acid-free paper

This Springer imprint is published by Springer Nature
The registered company is Springer International Publishing AG Switzerland

Preface

Many times in science we have witnessed the "rebirth" of old ideas and discoveries that were abandoned for a long time and even discarded! And then suddenly, one day they come back to stay and become very "trendy" subjects. A couple of examples quickly come to my mind: cancer immunotherapy and regulatory T cells. The crucial role of the immune system in controlling neoplasms was proposed more than a century ago. However, this has been widely accepted in the biomedical community only after the therapeutic success of, for example, immune checkpoint inhibitors. The case of regulatory T cells is even more compelling. A significant number of research groups during the 1970s and early 1980s described a particular subset of immunosuppressive T cells. A careful review of these early papers reveals that the experiments carried out with these suppressive T cells are surprisingly similar to the current trendy "Treg" experiments. Unfortunately, research on suppressive T cells abruptly stopped during the early 1980s due to the lack of specific markers identifying these cells. These cells were difficult to isolate, and the reproducibility of suppression assays was rather poor. Then, thanks to Sakaguchi and colleagues, work on suppressive T cells was strongly restarted after they identified natural regulatory T cells based on high expression of the CD25 marker. These cells could be isolated and worked with.

Could it be possible that research on myeloid-derived suppressor cells (MDSCs) is another example? It might be so. The pro-carcinogenic role of tumor-infiltrating myeloid cells was evident since as early as the 1970s. These myeloid cells were highly immature and lacked expression of other lineage markers. Then research on immunogenic cell lineages took over the study of these "obscure" pro-carcinogenic subsets. Everybody was studying macrophages, dendritic cells, neutrophils, eosinophils, and obviously, T cells and natural killer cells. Around the year 2000, a few groups identified these pro-carcinogenic myeloid cells by the expression (and lack of expression) of certain markers. Suddenly the research on MDSCs increased so much that the number of papers on MDSCs increased from about a dozen 10 years ago up to nearly 500 in 2015 alone.

We think that MDSC research has just started and we are experiencing a "rebirth" of an old subject, thanks to the pioneering work of a small number of groups. As MDSCs are relatively unknown (although this is quickly changing), the authors of this book thought that it was worthy to write a guide to the specialized reader who wants to know more about this myeloid subset.

We sincerely hope that we have achieved our goal of writing a concise but thorough review on the current knowledge on myeloid-derived suppressor cells.

David Escors

Contents

Chapter 1
Controversies in Neoplastic Myeloplasia

James E. Talmadge

Abstract The neutrophilia observed in cancer patients is associated with T-cell immunosuppression, disease progression, and a poor prognosis. In recent years, this has been reported to be due to the expansion of immature myelopoietic progenitors whose differentiation has been arrested and which are identified as myeloid-derived suppressor cells (MDSCs). However, despite the recent and intense focus on these cells, their phenotypes and their role in tumor progression remain controversial. In this chapter, we have focused on five of these controversies: (1) What are MDSCs phenotypically? (2) Is T-cell suppression by MDSCs antigen specific? (3) What are the differences between PMN-MDSCs and neutrophils (PMNs)?; (4) What are the differences between M-MDSCs and monocytes/macrophages?; and (5), What are the clinically effective therapeutic interventions for MDSCs? While there are other controversies in the MDSC realm, we suggest that these are currently the critical questions on which our understanding of their basic, translational and clinical importance.

Keywords Myeloplasia · MDSC · Neutrophil · Monocyte · Macrophage · Cancer

1.1 Introduction

In cancer patients, neutrophilia, leukoplasia, and monocytosis (leukemoid reaction) have been associated with T-cell immunosuppression and a poor prognosis [1]. The origin of these immunologic abnormalities and their contribution to disease pathogenesis remains a controversy, although our knowledge and understanding has advanced significantly. Within this chapter, we discuss a few of the controversies in the pathogenesis of leukemoid reactions, which we now know not only occurs with neoplasia, but also infectious diseases, inflammatory conditions, and autoimmunity. We have focused on several critical controversies and one appro-

J.E. Talmadge (✉)
Department of Pathology and Microbiology, University of Nebraska Medical Center,
68198-6495 Omaha, NE, USA
e-mail: jtalmadg@unmc.edu

© The Author(s) 2016
D. Escors et al., *Myeloid-Derived Suppressor Cells and Cancer*,
SpringerBriefs in Immunology, DOI 10.1007/978-3-319-26821-7_1

priate intervention to reduce myeloid-derived suppressor cell (MDSCs) numbers, providing an intellectual exercise that has not yet reached the level of controversy but rather an area warranting discussion, as we do not yet know the MDSC subtype to be targeted. Immunotherapy is now an attractive strategy for the treatment of neoplasia. However, there are impediments to the efficacy of the current approaches. One of these obstructions is the cells that blunt spontaneous or induced immune responses, including MDSCs, a group of pathologically activated immature myeloid cells with immunosuppressive capacity. MDSCs were initially described in the mid-1960s when tumors were reported to induce a leukemoid

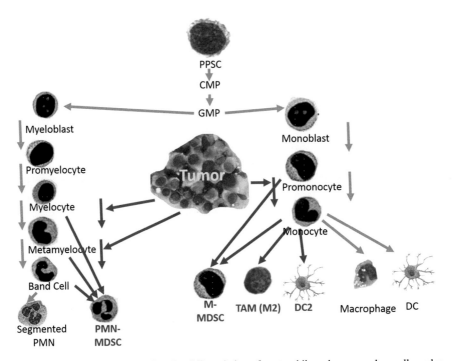

Fig. 1.1 MDSCs in myeloplasia. The differentiation of neutrophils and mononuclear cells under steady-state conditions is shown by the *blue arrows* in this graphic. On a simplistic base, hematopoietic stem cells (HSC) differentiate into common myeloid progenitors (CMP) and then committed granulocyte-macrophage progenitors (GMP). The GMP can differentiate into mature neutrophils via sequential steps including myeloblasts, promyelocytes, myelocytes, metamyelocytes, and bands or into macrophages and DCs similarly via sequential steps including monoblasts, promonocytes, and monocytes. The differentiation of myeloid cells under pathologic conditions (i.e., tumor-bearing mice) is shown by the *red arrows*. Tumor-derived growth factors affect all steps of granulocytic and monocytic cell differentiation and can result in the expansion of pathologically activated PMN-MDSCs and M-MDSCs. During tumor growth and other inflammation associated pathologies occur at a high frequency in the spleen and blood; although non-suppressive phenotypic counterparts maybe observed in the marrow of both normal and pathologic hosts in association with the hematopoietic progenitor characteristics of MDSCs. Further, M-MDSCs, at a primary or secondary tumor site can differentiate to TAMs and DC2s that are also immunosuppressive. In addition to the growth factor driven proliferation of MDSCs, their differentiation into mature myeloid is arrested, further facilitating their numbers

reaction [2, 3]. The expansion of myeloid cells including MDSCs is not only associated with tumor growth, but is also a major component of inflammatory and hematopoietic processes [4]. Subsequent studies revealed an increased cellularity in lymphoid and some parenchymal organs during tumor growth [3, 5] following Bacillus Chalmette-Guerin (BCG) injection [4, 6, 7]. However, it was not until 2007 that these myeloid cells were named MDSC and their murine phenotype was formally described [8]. In recent years, an increased understanding regarding the biology and clinical significance of MDSCs has been reported. In this short discussion, we expand on the unresolved issues associated with MDSCs. They can be broadly defined as immature myeloid cells that differ from terminally differentiated mature myeloid cells and include immature and progenitor myeloid cells; although, they are morphologically and phenotypically similar to monocytes (M-MDSCs), and polymorphonuclear (PMN) neutrophils (PMN-MDSCs) (Fig. 1.1).

1.2 Controversy 1: What Are MDSCs Phenotypically?

Phenotypic, functional, and morphologic heterogeneity are hallmarks of MDSCs. The plasticity of this myeloid compartment and the pathologic conditions that increase their numbers, as well as, the similar immunosuppressive activity independent of the mechanisms has resulted in an ambiguous definition of MDSC. Critically, they are defined functionally based on the inhibition of T-cell function and viability. In mice, the phenotype of MDSCs is $CD11b^+Gr1^+$ cells, which are subset by the variable expression of Gr-1 with $Gr-1^{hi}$ cells being identified as PMN-MDSCs and $Gr-1^{lo}$ cells identified as M-MDSCs [9]. The Gr-1 antibody (RB6-8C5) binds the same epitope as anti-Ly6G (IA8), such that cells stained with anti-Gr1 cannot be stained with anti-Ly6G. The Gr-1 antibody also stains cells that express Ly6C (ER-MP20 or AL-21) but not in a competitive manner [10, 11]. However, staining panels with anti-Ly6G and anti-Ly6C allow more accurate identification of M-MDSC ($CD11b^+Ly6C^{hi}Ly6G^-$ and PMN-MDSC ($CD11b^+Ly6C^{-/lo}Ly6G^+$) [12, 13]. Thus, there has been some confusion within the literature regarding staining and expression of Gr-1, as well as reports where investigators have stained using both anti-Gr-1 and anti Ly-6G antibodies.

In humans, MDSCs were originally described as $CD34^+$ hematopoietic progenitors with immune suppressive activity [14]. They are found in the mononuclear fraction following Ficoll Hypaque separation of blood as the PMN-MDSCs, which have a segmented neutrophil morphology and are hypodense [15, 16]. PMN-MDSCs are now identified as $CD14^-CD11b^+CD33^+$ and $CD15^+$ or $CD66b^+$ cells. It is noted that some investigators also include $HLA-DR^{-/lo}$ and/or Lin^- as markers. M-MDSCs are defined as $CD14^+HLA-DR^{-/lo}$; although, many investigators include as part of this phenotype Lin^- and $CD15/CD66b^-$, primarily because these markers are included in the staining tube. Confounding the identification of MDSCs is the use of the $Lin^-HLA-DR^-CD33^+$ phenotype, which incorporates a mixed group of cells that incorporate myeloid progenitors including MDSCs. One of the challenges to studying MDSCs is that their accurate characterization in cancer patients requires the phenotypic analysis of all three cell populations. Several other markers contribute to the characterization of MDSCs; however, none has emerged as a unique MDSC marker [17, 18]. One informative phenotypic marker that can be included in a staining panel is CD34 as a marker of the hematopoietic progenitor properties of MDSCs [14].

A reduction in MDSCs in tumor-bearing mice and cancer patients significantly improves immune responses and in some animal models and clinical studies results in antitumor activity [19, 20]. In addition, MDSC have a role in non-immunological functions, including the promotion of vasculogenesis, osteolysis and tumor cell invasion, and metastasis [21–24]. The original defining feature of MDSCs and myeloid cells with immunosuppressive function was their in vitro ability to suppress immune function, which remains the *sine qua none* for MDSCs.

The mechanisms of MDSC-mediated immune suppression include arginase (ARG1), inducible nitric oxide synthase (NOS) (iNOS) [9, 12, 13], TGF-β [25, 26], IL-10 [27] and COX2 [28, 29], sequestration of cysteine [30], decreased L-selectin expression by T cells [31], and induction/expansion of Tregs [32]. PMN-MDSCs and M-MDSCs have differing immunosuppressive mechanisms; with M-MDSCs utilizing mechanisms associated with NO and cytokines. In contrast, PMN-MDSCs suppress T-cell responses by reactive oxide synthase (ROS) production. The reaction of NO with superoxide generates peroxynitrite (PNT), which inhibits T-cells by nitrating T-cell receptors (TCRs), reducing their responsiveness to antigen-MHC complexes [33]. Additionally, nitration reduces binding of antigenic peptides to MHC molecules on tumor cells [34] and blocks T-cell migration by nitrating T-cell-specific chemokines [35]. S-nitrosylation also regulates cellular inflammation including nitrosylation of signaling molecules, integrins, and cytokines [36].

1.3 Controversy 2: Is T-Cell Suppression by MDSCs Antigen Specific?

A second controversy is the antigen-specific nature of MDSC-mediated immune suppression. The issue of antigen-specific suppression of T cells is important to understanding the biology of the immune defects in cancer patients. MDSC accumulation, with potent non-specific immune suppressive activity, is observed in the circulation and peripheral lymphoid organs, potentially resulting in systemic immune suppression. Indeed, this is a common observation that is associated with both a depression in T-cell numbers and function [37], limiting the host response to immunization [38]. A similar controversy exists in the results reported with blood obtained from cancer patients [39, 40]; although in most of these experiments, the specific nature of T-cell suppression has not been investigated [41]. This is a relevant comment as most clinical and preclinical studies of MDSC function use irrelevant (frequently alloantigens) to assess the suppression of T-cell function. This is not to suggest that antigen-specific suppression of T cells does not occur.

The major subset of MDSCs responsible for CD8$^+$ tolerance are G-MDSCs, due to their prevalence in the lymphoid organs of tumor-bearing hosts and their mechanism of immune suppression. G-MDSC have high levels of ROS and peroxynitrites (PNT) and ROS are short-lived and highly reactive, they function only close to T cells. As such, the interface of MDSC and CD8$^+$ T-cells interactions during antigen-TCR recognition is such an environment [33]. The functional activity of MDSC includes the inhibition of IFN-production by CD8$^+$ T cells, in response to peptide epitopes presented by major histocompatibility complex (MHC) class I in vitro and in vivo [42]. This antigen-specific T-cell tolerance depends on MHC class I, is not mediated by soluble factors, requires direct cell–cell contact, and is mediated by reactive oxygen species [43, 44].

M-MDSCs suppress both antigen-specific and non-specific T-cell responses and, on a per cell basis, may have a heightened suppressive activity relative to PMN-MDSCs [13, 41]. In contrast to M-MDSC, PMN-MDSCs predominantly suppress T-cell responses in an antigen-specific manner [13]. Different effects of MDSC on T-cell responses in cancer patients and tumor-bearing mice have been reported [41]. It has been suggested that MDSC induce antigen-specific tolerance of CD8$^+$, but not CD4$^+$ T cells [43–45]. However, in some model systems, MDSC mediate the inhibition of IFN-γ production by CD4$^+$ T cells [46–48]. The ability of MDSC to induce antigen-specific CD4$^+$ T-cell tolerance in vivo may be dependent on MHC class II expression [49]. In most tumor models that have been examined, the expression of MHC class II molecules, on MDSC, was significantly lower than on myeloid cells with the same phenotypic differentiation in tumor-free mice. Indeed, in humans the lack of expression of HLA-DR is a defining characteristic of MDSCs.

1.4 Controversy 3: What Are the Differences Between PMN-MDSCs and Neutrophils (PMNs)?

Neutrophils with immunosuppressive and pro-tumorigenic activity are identified as N2 neutrophils, as opposed to antitumor N1 neutrophils in some reports [50, 51]. However, as these are short-lived, terminally differentiated cells (PMNs) it has been suggested that N1-cells represent bona fide, activated PMN cells, whereas N2-cells are PMN-MDSCs. However, this conundrum cannot be resolved without markers that allow the delineation of PMN cells versus PMN-MDSCs. In mice, several markers have been identified that can distinguish PMN-MDSCs from PMNs have been suggested [52]; however, multiple markers are required to differentiate between the two cell types.

In humans, neutrophils can be separated from PMN-MDSCs and M-MDSCs by density gradients or elutriation, with MDSCs separating in the mononuclear fraction (hypodense) while neutrophils are separated in the denser fraction [15, 53]. In normal donors, PMN-MDSCs are rare in the peripheral blood mononuclear cell

(PBMC) fraction and while useful, cell density as a clinical biomarker is limited as the density of PMNs is dependent on the blood collection conditions, storage, and activation [54]. One approach that we and others have used to differentiate human PMN from PMN-MDSCs is the co-expression of CD16 and CD11b. Both of these markers are absent on promyelocytes and appear during neutrophil differentiation at the myelocyte/metamyelocyte stage. During this process, expression of CD11b precedes expression of CD16 [55]. When mature PMNs express both CD11b and CD16, while CD16 expression is low or negative on immature myeloid cells providing a potential marker to separate PMNs from PMN-MDSCs [56].

1.5 Controversy 4: What Are the Differences Between M-MDSCs and Monocytes/Macrophages?

Phenotypic identification of M-MDSCs is similarly challenging, although it is less critical in humans relative to mice as monocytes can be distinguished from M-MDSCs by their phenotype (CD14$^+$HLA-DR^{hi} vs. CD14$^+$HLA-DR$^{-/lo}$, respectively). However, M-MDSC infiltrating human tumors can be more challenging. Immature M-MDSCs (as well as immature PMN-MDSCs) are CD34$^+$ and CD117$^+$. Both the immature and "mature" MDSCs are CD14$^+$ and HLA-DR$^{lo/-}$ allowing differentiation from monocytes, macrophages, and PMN-MDSCs. However, human tumor-associated macrophages (TAM) are subdivided into M1-like and M2-like cells. The M1-like TAM are CD64$^+$CD80$^+$ and CXCR10$^+$ while the M2-like TAM are CD32$^+$CD163$^+$CD23$^+$CD200R$^+$PD-L2$^+$. Classically activated (or M1) macrophages are elicited in an environment dominated by Th1 cytokines, such as IFN-γ and TNF-α, and/or by recognition of pathogen-associated molecular patterns or endogenous danger signals. This is a prototypical pro-inflammatory type of macrophage that is implicated in the initiation and propagation of inflammation and pathogen clearance. Macrophage functions and phenotypes are altered by Th2 cytokines resulting in alternatively activated macrophages or M2 macrophages [57]. A feature common to M2 macrophages is their ability to suppress Th1 cytokine-driven inflammation and regulate adaptive immune responses, i.e., T-cell immunosuppressive activity similar to MDSCs that has resulted in confusion in the literature. This has been exacerbated by the suggestion that M-MDSC may express both M1 and M2 precursor phenotypes such that a M2 prototypic M-MDSC can be measured based on immunoglobulin-like transcript 3 (ILT3) membrane expression [58]. In the tumor microenvironment, MDSC with the M2-like phenotype are dominant and produce large amounts of IL-10 and arginase, induce anergy of antitumor immune cells, and expand immunosuppressive regulatory T cells (Treg). Indeed, separation of non-small cell lung cancer patients on the basis of ILT3 membrane expression on MDSCs, i.e., ILT3^{low} and ILT3^{high} populations revealed that patients with an increased frequency of ILT3^{high} MDSCs had a shorter median survival than patients with an increased frequency of ILT3^{lo} MDSCs [58].

1.6 Controversy 5: What Are Clinically Effective Therapeutic Interventions for MDSCs?

Recent reports have documented that the frequency and absolute number of MDSCs in peripheral blood of cancer patients and spleens of tumor-bearing mice directly correlate with tumor burden and a worse clinical outcome [59] and inversely with T-cell number/function [60]. The regulation of MDSC proliferation and function incorporates a complex network of molecular pathways and therefore, many different therapeutic strategies are being developed to regulate them and restore host immunity. Controlling MDSC-mediated immune dysfunctions can be achieved by one or more of four approaches (Table 1.1): (1) inhibition of immune suppressive activity; (2) reducing their number; (3) regulating their development/apoptosis or by; and (4) inducing their differentiation.

1.6.1 Inhibiting MDSC Immunosuppressive Activity

One approach to constrain MDSC activity is to target their immune regulatory functions. For example, the selective inhibitors of the janus kinase and signal transducer and activator of transcription (JAK2/STAT3) pathway, by cucurbitacin B in advanced lung cancer patients, significantly reduced the Lin⁻HLA-DR⁻CD33⁺ immature MDSCs in the peripheral blood [61]. Phosphodiesterase-5 (PDE-5) inhibitors, by increasing intracellular cyclic guanosine monophosphate (cGMP) levels have been found to decrease IL-4Rα expression, and downregulate iNOS levels in intratumoral MDSCs, reversing MDSC immunosuppressive activity and increasing T-cell function in vivo [62]. In a clinical study (number NCT00894413), tadalafil in patients with head and neck squamous cell carcinoma documented a significant reduction in ARG1 and iNOS activity. Moreover, tadalafil also decreased CD33⁺HLA-DR⁻IL-4Rα⁺ MO-MDSCs in both the blood and tumors of treated patients [63]. In line with these results, the administration of tadalafil to a patient with end-stage relapsed/refractory multiple myeloma reduced MDSC function and associated with a durable anti-myeloma immune and clinical response [64]. Moreover, a clinical trial (NCT01374217) prospectively evaluating the effect of MDSC inhibition in myeloma will analyze patients refractory to lenalidomide-based treatments receiving tadalafil in addition to their lenalidomide-containing regimen; the study, which is not yet concluded will provide indications whether tadalafil could improve the outcome for patients pretreated with lenalidomide by decreasing MDSC frequency. Very recently, tadalafil treatment (clinical trial number NCT008436359) significantly reduced both CD33⁺HLA-DR⁻IL-4Rα⁺ M-MDSCs and Tregs in the blood and in tumor of HNSCC patients and increased tumor-specific CD8⁺ T cells [65]. The mechanisms by which tadalafil contributes to affect MDSC activity are not clear: PDE5 inhibition might decrease IL-4Rα expression on MDSCs, reducing the survival signaling provided by the receptor [66]. However, the role of IL-4Rα

Table 1.1 Interventions targeting MDSC expansion, trafficking, and activation

Therapeutic agent	Type of cancer tested	Effect on MDSCs	References
COX2 inhibitor (Celebrex)	Mammary carcinoma (mice), ovarian cancer, and melanoma (human)	Inhibition of proliferation	[29, 28, 112]
Amino-biphosphonate	Mammary tumors (mice) Pancreatic adenocarcinoma (human)	Inhibition of proliferation	[85, 87]
Phosphodiesterase-5 inhibitor (sildenafil and tadalafil)	Mammary carcinoma, colon carcinoma, and fibrosarcoma (all mice) head and neck (human)	Inhibition of proliferation and of suppressive effects	[62, 63, 65]
c-KIT-specific antibody	Colon carcinoma (mice)	Inhibition of proliferation	[98]
Nitroaspirin	Colon carcinoma (mice & human)	Inhibition of suppressive effects	[68]
Triterpenoid (CDDO-Me)	Colon carcinoma, thymoma, and lung cancer (mice), pancreatic adenocarcinoma (human)	Inhibition of suppressive effects	[70]
All-trans retinoic acid	Sarcoma and colon carcinoma (mice) Metastatic renal cell carcinoma and lung cancer (human)	Inhibition of proliferation	[108, 110, 113]
25-hydroxyvitamin D3	Lewis lung carcinoma (mice) and head and neck and lung cancer (human)	Moderate inhibition of proliferation and induction of differentiation	[114–116]
Paclitaxel	Melanoma (mice)	Matures MDSCs into DCs	[117]
Gemcitabine	Lung and breast cancer (mice), pancreatic adenocarcinoma	Inhibition of proliferation	[75, 81, 82, 118]
VEGF-trap	Solid tumors (human)	No activity	[104]
Antibody to VEGFA (bevacizumab)	Lung, breast, colorectal carcinoma, and metastatic renal cell cancer (human)	Weak inhibition of proliferation	[39, 105, 119]

(continued)

Table 1.1 (continued)

Therapeutic agent	Type of cancer tested	Effect on MDSCs	References
Doxorubicin-cyclophosphamide	Breast cancer (human)	Weak inhibition of proliferation	[79]
CXCR2 (S-265610) and CXCR4 (AMD3100) antagonists	Mammary cancer (mice)	Altered recruitment of immature myeloid cells to the tumor	[120]
Tyrosine kinase inhibitor (sunitinib)	Renal cell cancer (human)	Weak inhibition of proliferation	[121]
PROK2-specific antibody	Various tumors of human and mouse origin in nude mice	Inhibition of polymorphonuclear MDSC expansion and recruitment	[122]
Neutralizing antibody to GM-CSF	Pancreatic cancer (mice)	Inhibit proliferation	[97]
Tripeptide CDDO-Me (RTA-402)	Pancreatic adenocarcinoma (human)	Inhibit proliferation	[70]
Neutralizing antibody to G-CSF	Colon carcinoma (mice)	Inhibit proliferation	[123, 124]
AT38 (an NO donor based on the furoxan molecule)	Fibrosarcoma and thymoma	Downregulation of ARG1, iNOS, and peroxynitrite in MDSCs; expression of nitrated or nitrosylated CCL2	[35]
CSF1R and KIT receptor tyrosine kinase inhibitor (PLX3397)	Mammary carcinoma (mice), melanoma, AML, and breast cancer (human)	Inhibition of TAM recruitment	[125]
CSF1R antagonist (GW2580)	Lung carcinoma and prostate cancer (human)	Inhibition of the expansion of MDSC and macrophage populations	[23]
5-fluorouracil	Thymoma	Inhibition of MDSC population expansion	[76]
Docetaxel	Mammary carcinoma (mice)	Inhibition of MDSC population expansion; macrophage polarization to M1 phenotype	[47]
Very small size proteoliposomes	Lymphomas and sarcoma	iNOS downregulation and changes in MDSC subset distribution	[126]

(continued)

Table 1.1 (continued)

Therapeutic agent	Type of cancer tested	Effect on MDSCs	References
Neutralizing antibody to CCL2	Mammary carcinoma (mice)	Targeting inflammatory monocytes and macrophages	[127]
IL-12	Melanoma and colon carcinoma	Differentiation of MDSCs, TAM, and DCs	[128, 129]
Ab to CD40 and IL2	Renal cell carcinoma (human)	Increased iNOS and TIMP1 and differentiation of TAM	[130]
CD40 agnostic Abs and gemcitabine	Pancreatic ductal adenocarcinoma (human)	Differentiation of TAM	[131]
Histidine-rich glycoprotein (HRG)	Fibrosarcoma, pancreatic cancer, and breast cancer	Down regulation of PIGF and differentiation of TAM	[132]
Inhibition of NF-kB signaling via IkB kinase	Ovarian cancer	Differentiation of TAM	[133]
Peroxisome proliferator-activated receptor-g (PPAR) inhibitor	Carcinoma and myeloid sarcoma	Induces proliferation of MDSCs	[134]
antibody to IL-6 and gemcitabine	Carcinoma	Inhibits accumulation of M and PMN-MDSCs	[135]
Curcubitacin B	Lung cancer	Inhibition of MDSC	[61]
Antibody to CSF1R (RG7155 & IMC CS4)	Advanced solid tumors	Inhibition of M-MDSC	[136]
Vemurafenib (B-RAF inhibitor)	Metastatic melanoma	Decrease PMN and M-MDSC	[106]
Ipilimumab (anti-CTLA-4)	Metastatic melanoma	Decrease M-MDSC	[137]
Anti-PD-L1	Multiple myeloma (mice)	Decrease MDSCs	[138]

Ab antibody; *ARG* arginase; *B-RAF rapidly accelerated fibrosarcomas*; *COX2* cyclooxygenase 2; *CSF1R* colony-stimulating factor 1 receptor; *CCL2* Chemokine Ligand 2; *CDDO-ME* Methyl-2-cyano-3,12-dioxooleana-1,9(11)-dien-28-oate; *CTLA-4* cytotoxic T-lymphocyte-associated protein; *CXCR* CXC chemokine receptor; *DC* dendritic cell; *G-CSF* granulocyte-colony-stimulating factor; *GM-CSF* granulocyte-macrophage colony-stimulating factor; *G-MDSCs* granulocyte-MDSCs; *HRG* Histidine-rich Glycoprotein; *IkB* inhibitor kappa beta; *iMCs* immature myeloid cells; *iNOS* inducible Nitric Oxide Synthase; *IIL* interleukin; MDSC, *c-KIT* stem cell factor receptor; myeloid-derived suppressor cell; *M-MDSCs* monocyte-MDSCs; *NF-kB* nuclear factor-kappa beta; *NO* nitric oxide; *PD-L1* programed death ligand; *PMN* polymorphonuclear; *PPARγ* peroxisome proliferator-activated receptor-γ; *PIGF* phosphatidylinositol glycan anchor biosynthesis class F; *PROK2* prokineticin 2; *TAM* tumour-associated macrophage; *TIMP* tissue inhibitor of metalloproteinase; *VEGFA* vascular endothelial growth factor A remains to be fully elucidated since, in some cases, this cytokine receptor does not seem to be involved in the immune suppressive properties of MDSCs [67]. Another possible explanation of the dramatic effect of tadalafil on MDSCs could be attributed to a destabilization of iNOS mRNA resulting in a reduction in NO production [63].

NO has a pivotal role in MDSC-dependent immune suppression and drugs that inhibit NOS are currently under investigation. Nitroaspirin, which inhibits iNOS, has been shown to normalize the immune status of tumor-bearing hosts and improve tumor-antigen-specific T-cell responses to cancer vaccination [68]. AT38, [3-(aminocarbonyl)furoxan-4-yl] methylsalicylate, an NO-donating compound, can decrease MDSC inhibitory activity by reducing CCL2 chemokine nitration promoting T cell infiltration of primary tumors, as well as, significantly reduce iNOS and ARG1 enzyme activity in myeloid cells and control RNS generation within the tumor environment [35].

Synthetic triterpenoids including CDDO-Me can induce the nuclear factor (erythroid-derived 2)-like 2 (NRF2)-mediated upregulation of antioxidant genes, including NAD(P)H quinone oxidoreductase 1 (NQO1), thioredoxin, catalase, superoxide dismutase, and heme oxygenase, reducing intracellular ROS levels and controlling the immune suppressive activity of MDSCs and improving host immune responses [69]. Clinical support for this strategy has been provided in pancreatic cancer patients with CDDO-Me in a phase I clinical trial (RTA 402-C-0702) that has shown a significant improvement in the host's immune response without reducing MDSC frequency in the patient's peripheral blood [70]. Since CDDO-Me might also have a role in STAT3 inhibition both in tumor and myeloid cells, in addition to ROS elimination, it represents an attractive therapeutic option to increase cancer immunotherapy.

Upregulation of COX2 levels is important to tumor progression and has been reported to correlate with a poor prognosis in multiple tumor pathologies [71]. Since COX2-derived prostaglandin E2 has a role in inducing ARG1 upregulation in MDSCs [72], COX2-based therapeutic approaches have been assessed in order to limit cancer progression. Dietary administration of the COX2 inhibitor celecoxib in mesothelioma-bearing mice induced a significant decrease in the number and suppressive functions of ROS-producing PMN-MDSCs both in the spleen and tumor microenvironment. Accordingly, celecoxib supplied in combination with DCs pulsed with tumor lysates improved the survival of mesothelioma-bearing mice compared with the single treatments [73].

1.6.2 Depleting MDSC Number

The antitumor activity of chemotherapy may also depend on off-target effects, most notably immunoregulatory activity contributing to successful tumor control [74]. Indeed, some conventional chemotherapy agents, including gemcitabine [75] and 5-fluouracil [76], have a cytotoxic activity on MDSCs. Cisplatin has also been reported to modify myeloid cell frequency by reducing MDSCs, potentially by inducing their differentiation into DCs and promoting antitumor T-cell responses in melanoma-bearing mice [77]. Paclitaxel, an inhibitor of microtubule disassembly, enforces MDSC differentiation to DCs [78], whereas docetaxel, a drug with similar action, selectively kills and polarizes MDSCs towards an M1-like phenotype [47]. Cyclophosphamide (CTX) may have conflicting effects on MDSCs, since in combination with doxorubicin it boosts MDSC frequency in breast cancer patients [79] but in combination with gemcitabine, mitigates Treg- and MDSC-mediated immunosuppression, triggering antitumor immunity in vivo [80]. It should be noted that in the first study, the patients were also injected with G-CSF, a growth factor for myeloid progenitors that has been shown to increase MDSC numbers clinically [79].

The combination of a personalized peptide vaccine with gemcitabine administration was shown to result in an additive effect, boosting immune cellular and humoral responses in unresectable pancreatic cancer patients [81]. In another study, a significant reduction in MDSCs (Lin⁻HLA-DR⁻CD11b⁺), in patients with advanced pancreatic cancer was observed following combination therapy with gemcitabine and capecitabine together with a GV1001 vaccine and GM-CSF as adjuvant [82]. The adjuvant activity of MDSC control was also shown with a reduction in the numbers of peripheral MDSCs in breast cancer patients treated with oxidized glutathione mimetic (NOV-002) and chemotherapy with doxorubicin and cyclophosphamide, resulting in a complete clinical response in patient with a lower frequency than in patients who did have a positive clinical outcome [83].

1.6.3   Regulating MDSC Development/Apoptosis

Bisphosphonates (BPs) are potent inhibitors of osteoclast-mediated bone resorption by the inhibition of farnesyl pyrophosphate synthase activity, a key regulatory enzyme in the mevalonic acid pathway critical to the production of sterols and isoprenoid lipids. As a result, the posttranslational modification (isoprenylation) of many proteins is inhibited, leading to osteoclast apoptosis [84]. Treatment with amino-bisphosphonates slows tumor growth, significantly decreasing MMP9 levels and macrophage numbers in tumor stroma, as well as reducing MDSC proliferation in the bone marrow and peripheral blood [85]. Therapeutic activity in mice bearing resected pancreatic adenocarcinomas was demonstrated with zoledronic acid (ZA), which significantly decreased Gr⁻1⁺CD11b⁺ MDSC accumulation in tumors, impaired tumor growth, and prolonged survival [86]. Based on these and other results, a phase 1 study (NCT00892242) to evaluate the safety and the efficacy of perioperative ZA administration in 23 patients with resectable pancreatic adeno-carcinoma was undertaken, but was not shown to reduce PMN-MDSCs [87]. Nonetheless, in three clinical trials of breast cancer patients adjuvant administration of zoledronic acid documented clinical activity. In the ABCSG-12 trial with 1803 patients, disease free survival at 62 months was increased from 88 to 92 %

(P = 0.009) with the addition of zoledronic acid to endocrine therapy [88]. In the ZOFAST study, zoledronic acid administration was given either immediately or delayed (after fracture or high risk thereof) administration with adjuvant endocrine therapy. The disease free survival increased by immediate zoledronic acid administration from 92 to 95 % (P = 0.0314), at 36 months follow-up [89]. In the AZURE trial, however, with 3360 patients, disease free survival was 77 % and no difference between zoledronic acid treatment and control was seen at a median follow-up of 59 months (P = 0.79) [90, 91]. In this study, the majority of patients received chemotherapy rather than endocrine therapy alone. A subgroup analysis in patients being postmenopausal for more than 5 years showed an increase in disease free survival from 71 to 78.2 % (P = 0.02) 5 years after randomization. A meta-analysis among 17,751 breast cancer patients from 41 randomized clinical trials compared outcomes with and without adjuvant bisphosphonate treatment and found reduction of breast cancer mortality and bone recurrence in postmenopausal patients [91]. Together, these results suggest that ZA can increase disease free survival of breast cancer patients, although the mechanism of action and the role of MDSC regulation are not clear [92].

S100A8/S100A9 proteins, together with their receptor for advanced glycation end products (RAGE), are involved in MDSC recruitment and retention [93]. Accordingly, mAbGB3.1, an anti-carboxylate glycan antibody that blocks S100A8/A9 binding and signaling, has been shown to reduce the serum levels of these proteins and consequently MDSC accumulation in blood and secondary lymphoid organs of tumor-bearing mice [94]. The inhibition of CSF-1R signaling through CSF1R kinase inhibitors is another strategy to regulate MDSCs. GW2580 has been shown to abrogate tumor recruitment of M-MDSCs in mice bearing Lewis Lung Carcinoma and also to decrease proangiogenic and immunosuppressive genes [23]. The relevance of blocking CSF-1/CSF-1R signaling was supported using a tumor model in which radiotherapy was followed by an increase in CSF-1 expression in mice bearing prostate cancer. In this tumor therapy model, CSF-1R inhibition with GW2580 or PLX3397 regulated myeloid infiltration and significantly delayed tumor regrowth post irradiation [95]. Currently, a monoclonal antibody against CSF-1R (IMC-CS4) is being studied in a phase I clinical trial (NCT01346358) to establish its safety and pharmacokinetic profile in the treatment of subjects with advanced solid tumors, either refractory to standard therapy or for whom no standard of care is available. Moreover, PLX-3397 is being studied in an ongoing phase I/II clinical trials with solid and hematological tumors. In preclinical studies, therapeutic activity has been observed with antibodies to G-CSF [96], GM-CSF [97] and SCF [98] in tumor-bearing mice. In the clinic, GM-CSF has been used as an adjuvant, to mature DCs ex vivo and as the transfected component of cell-based vaccines. However, high doses of GM-CSF induce the accumulation of MDSC in mouse systems [99, 100], and studies in which tumor production of GM-CSF was knocked down by RNA interference demonstrated that GM-CSF expands MDSC [9]. A clinical trial in stage IV metastatic melanoma patients demonstrated that a GM-CSF-based vaccine induced MDSCs that suppressed immunity [101]. Therefore, although GM-CSF matures DC, it also induces MDSC, suggesting a need for close attention to the dosing of GM-CSF to promote DC function without inducing immune suppression.

Anti-angiogenic therapies have also been evaluated for their impact on MDSCs [102]. Sunitinib is a multi-kinase inhibitor of flt3, VEGFR, and c-kit signaling that has shown regulatory activity for MDSCs. Renal cell carcinoma (RCC) patients treated with sunitinib have documented a dramatic reduction in M-MDSCs, but no correlation between MDSC levels and tumor burden [103]. In a study of 15 patients with refractory solid tumors enrolled in a phase I clinical trial and treated with VEGF-Trap, no decrease in peripheral MDSC levels was reported [104]. In contrast, a study with lung, breast, and colorectal cancer patients reported a decrease in MDSCs and a slight increase in DC frequency following one cycle of bevacizumab, a monoclonal antibody directed against VEGFA [105]. Treatment of melanoma patients with vemurafenib, a specific inhibitor of mutant B-RAF V600E, a mutation leading to constitutive activation of the MAP kinase pathway, reduced the frequency of both M- and PMN-MDSCs in patients with cutaneous melanoma [106].

1.6.4   Inducing MDSC Differentiation

Promoting differentiation of suppressive MDSCs into mature, non-suppressive cells has been investigated in preclinical and clinical cancer models see review in [107]. All-trans retinoic acid (ATRA) is one of the first differentiating agents tested for MDSCs. The stimulation of myeloid maturation and the advantage induced by ATRA treatment was initially documented in mouse models, in which ATRA administration was associated with an increased efficacy of cancer vaccination [108]. Moreover, ATRA supplemented with GM-CSF induced the upregulation of HLA-DR expression, CD1a and CD40 on MDSCs isolated from RCC patients [109]. A randomized phase II clinical trial tested whether ATRA treatment of SCLC patients could increase the efficacy of chemotherapy associated with a vaccine consisting of DCs transduced with P53-expressing adenoviral vector; a significant reduction in MDSCs was only observed in patients treated with vaccination in combination with ATRA [110]. Similar to ATRA, vitamins D3 could also act as differentiating agents, as observed in two cohorts of newly diagnosed HNSCC patients who did or did not receive 125(OH)2 D3 vitamin treatment for 3 weeks before surgical resection. Only treated patients showed a reduction in intratumoral $CD34^+$ myeloid cells, paralleled by increased amounts of intratumoral mature DCs [111].

References

1. Gary W, Wood JEN, Stephens R (1979) Relationship between monocytosis and T-lymphocyte function in human cancer. Natl Cancer Inst 63(3):587–592
2. Lappat EJ, Cawein M (1964) A study of the leukemoid response to transplantable a-280 tumor in mice. Cancer Res 24:302–311
3. Lee MY, Rosse C (1982) Depletion of lymphocyte subpopulations in primary and secondary lymphoid organs of mice by a transplanted granulocytosis-inducing mammary carcinoma. Cancer Res 42(4):1255–1260
4. Bennett JA, Rao VS, Mitchell MS (1978) Systemic bacillus Calmette-Guerin (BCG) activates natural suppressor cells. Proc Natl Acad Sci USA 75(10):5142–5144
5. Tsuchiya Y, Igarashi M, Suzuki R, Kumagai K (1988) Production of colony-stimulating factor by tumor cells and the factor-mediated induction of suppressor cells. J Immunol 141 (2):699–708
6. Bennett JA, Mitchell MS (1979) Induction of suppressor cells by intravenous administration of Bacillus Calmette-Guerin and its modulation by cyclophosphamide. Biochem Pharmacol 28(12):1947–1952
7. Wren SM, Wepsic HT, Larson CH, De Silva MA, Mizushima Y (1983) Inhibition of the graft-versus-host response by BCGcw-induced suppressor cells or prostaglandin E1. Cell Immunol 76(2):361–371
8. Gabrilovich DI, Bronte V, Chen SH, Colombo MP, Ochoa A, Ostrand-Rosenberg S, Schreiber H (2007) The terminology issue for myeloid-derived suppressor cells. Cancer Res 67(1):425; author reply 426. doi:10.1158/0008-5472.CAN-06-3037
9. Dolcetti L, Peranzoni E, Ugel S, Marigo I, Fernandez Gomez A, Mesa C, Geilich M, Winkels G, Traggiai E, Casati A, Grassi F, Bronte V (2010) Hierarchy of immunosuppressive strength among myeloid-derived suppressor cell subsets is determined by GM-CSF. Eur J Immunol 40(1):22–35. doi:10.1002/eji.200939903
10. Ribechini E, Leenen PJ, Lutz MB (2009) Gr-1 antibody induces STAT signaling, macrophage marker expression and abrogation of myeloid-derived suppressor cell activity in BM cells. Eur J Immunol 39(12):3538–3551. doi:10.1002/eji.200939530
11. Rose S, Misharin A, Perlman H (2012) A novel Ly6C/Ly6G-based strategy to analyze the mouse splenic myeloid compartment. Cytometry Part A J Int Soc Anal Cytol 81(4):343–350. doi:10.1002/cyto.a.22012
12. Youn JI, Nagaraj S, Collazo M, Gabrilovich DI (2008) Subsets of myeloid-derived suppressor cells in tumor-bearing mice. J Immunol 181(8):5791–5802
13. Movahedi K, Guilliams M, Van den Bossche J, Van den Bergh R, Gysemans C, Beschin A, De Baetselier P, Van Ginderachter JA (2008) Identification of discrete tumor-induced myeloid-derived suppressor cell subpopulations with distinct T cell-suppressive activity. Blood 111(8):4233–4244. doi:10.1182/blood-2007-07-099226
14. Pak AS, Wright MA, Matthews JP, Collins SL, Petruzzelli GJ, Young MR (1995) Mechanisms of immune suppression in patients with head and neck cancer: presence of CD34(+) cells which suppress immune functions within cancers that secrete granulocyte-macrophage colony-stimulating factor. Clin Cancer Res Official J Am Assoc Cancer Res 1(1):95–103
15. LaFace D, Talmadge J (2011) Meeting report: regulatory myeloid cells. Int Immunopharmacol 11(7):780–782. doi:10.1016/j.intimp.2011.01.031
16. Ko J, Rayman Patricia, Obery Dana, Lindner Daniel, Borden Ernest, Finke James (2013) Proangiogenic neutrophilic-myeloid-derived suppressor cells emerge via two parallel pathways in renal cell carcinoma and melanoma. J Immunother Cancer 1:159
17. Talmadge JE, Gabrilovich DI (2013) History of myeloid-derived suppressor cells. Nat Rev Cancer 13(10):739–752. doi:10.1038/nrc3581

18. Damuzzo V, Pinton L, Desantis G, Solito S, Marigo I, Bronte V, Mandruzzato S (2015) Complexity and challenges in defining myeloid-derived suppressor cells. Cytometry Part B Clin Cytometry 88(2):77–91. doi:10.1002/cyto.b.21206
19. Abe F, Younos I, Westphal S, Samson H, Scholar E, Dafferner A, Hoke TA, Talmadge JE (2010) Therapeutic activity of sunitinib for Her2/neu induced mammary cancer in FVB mice. Int Immunopharmacol 10(1):140–145. doi:10.1016/j.intimp.2009.09.023
20. Ostrand-Rosenberg S, Sinha P (2009) Myeloid-derived suppressor cells: linking inflammation and cancer. J Immunol 182(8):4499–4506. doi:10.4049/jimmunol.0802740
21. Talmadge JE, Fidler IJ (2010) AACR centennial series: the biology of cancer metastasis: historical perspective. Cancer Res 70(14):5649–5669. doi:10.1158/0008-5472.CAN-10-1040
22. Ahn GO, Brown JM (2008) Matrix metalloproteinase-9 is required for tumor vasculogenesis but not for angiogenesis: role of bone marrow-derived myelomonocytic cells. Cancer Cell 13 (3):193–205. doi:10.1016/j.ccr.2007.11.032
23. Priceman SJ, Sung JL, Shaposhnik Z, Burton JB, Torres-Collado AX, Moughon DL, Johnson M, Lusis AJ, Cohen DA, Iruela-Arispe ML, Wu L (2010) Targeting distinct tumor-infiltrating myeloid cells by inhibiting CSF-1 receptor: combating tumor evasion of antiangiogenic therapy. Blood 115(7):1461–1471. doi:10.1182/blood-2009-08-237412
24. Sawant A, Deshane J, Jules J, Lee CM, Harris BA, Feng X, Ponnazhagan S (2013) Myeloid-derived suppressor cells function as novel osteoclast progenitors enhancing bone loss in breast cancer. Cancer Res 73(2):672–682. doi:10.1158/0008-5472.CAN-12-2202
25. Huang B, Pan PY, Li Q, Sato AI, Levy DE, Bromberg J, Divino CM, Chen SH (2006) Gr-1+ CD115+ immature myeloid suppressor cells mediate the development of tumor-induced T regulatory cells and T-cell anergy in tumor-bearing host. Cancer Res 66(2):1123–1131. doi:10.1158/0008-5472.CAN-05-1299
26. Li H, Han Y, Guo Q, Zhang M, Cao X (2009) Cancer-expanded myeloid-derived suppressor cells induce anergy of NK cells through membrane-bound TGF-beta 1. J Immunol 182 (1):240–249
27. Mundy-Bosse BL, Young GS, Bauer T, Binkley E, Bloomston M, Bill MA, Bekaii-Saab T, Carson WE 3rd, Lesinski GB (2011) Distinct myeloid suppressor cell subsets correlate with plasma IL-6 and IL-10 and reduced interferon-alpha signaling in CD4(+) T cells from patients with GI malignancy. Cancer Immunol Immunother CII 60(9):1269–1279. doi:10. 1007/s00262-011-1029-z
28. Mao Y, Poschke I, Wennerberg E, Pico de Coana Y, Egyhazi Brage S, Schultz I, Hansson J, Masucci G, Lundqvist A, Kiessling R (2013) Melanoma-educated CD14+ cells acquire a myeloid-derived suppressor cell phenotype through COX-2-dependent mechanisms. Cancer Res 73(13):3877–3887. doi:10.1158/0008-5472.CAN-12-4115
29. Talmadge JE, Hood KC, Zobel LC, Shafer LR, Coles M, Toth B (2007) Chemoprevention by cyclooxygenase-2 inhibition reduces immature myeloid suppressor cell expansion. Int Immunopharmacol 7(2):140–151. doi:10.1016/j.intimp.2006.09.021
30. Srivastava MK, Sinha P, Clements VK, Rodriguez P, Ostrand-Rosenberg S (2010) Myeloid-derived suppressor cells inhibit T-cell activation by depleting cystine and cysteine. Cancer Res 70(1):68–77. doi:10.1158/0008-5472.CAN-09-2587
31. Hanson EM, Clements VK, Sinha P, Ilkovitch D, Ostrand-Rosenberg S (2009) Myeloid-derived suppressor cells down-regulate L-selectin expression on CD4+ and CD8[+] T cells. J Immunol 183(2):937–944. doi:10.4049/jimmunol.0804253
32. Hoechst B, Ormandy LA, Ballmaier M, Lehner F, Kruger C, Manns MP, Greten TF, Korangy F (2008) A new population of myeloid-derived suppressor cells in hepatocellular carcinoma patients induces CD4(+)CD25(+)Foxp3(+) T cells. Gastroenterology 135(1):234–243. doi:10.1053/j.gastro.2008.03.020
33. Nagaraj S, Gupta K, Pisarev V, Kinarsky L, Sherman S, Kang L, Herber DL, Schneck J, Gabrilovich DI (2007) Altered recognition of antigen is a mechanism of CD8[+] T cell tolerance in cancer. Nat Med 13(7):828–835. doi:10.1038/nm1609

34. Lu T, Ramakrishnan R, Altiok S, Youn JI, Cheng P, Celis E, Pisarev V, Sherman S, Sporn MB, Gabrilovich D (2011) Tumor-infiltrating myeloid cells induce tumor cell resistance to cytotoxic T cells in mice. J Clin Investig 121(10):4015–4029. doi:10.1172/JCI45862

35. Molon B, Ugel S, Del Pozzo F, Soldani C, Zilio S, Avella D, De Palma A, Mauri P, Monegal A, Rescigno M, Savino B, Colombo P, Jonjic N, Pecanic S, Lazzarato L, Fruttero R, Gasco A, Bronte V, Viola A (2011) Chemokine nitration prevents intratumoral infiltration of antigen-specific T cells. J Exp Med 208(10):1949–1962. doi:10.1084/jem.20101956

36. Bhopale VM, Yang M, Yu K, Thom SR (2015) Factors associated with nitric oxide-mediated beta2 integrin inhibition of neutrophils. J Biol Chem 290(28):17474–17484. doi:10.1074/jbc.M115.651620

37. Singh RK, Varney ML, Buyukberber S, Ino K, Ageitos AG, Reed E, Tarantolo S, Talmadge JE (1999) Fas-FasL-mediated CD4+ T-cell apoptosis following stem cell transplantation. Cancer Res 59(13):3107–3111

38. Chen S, Akbar SM, Miyake T, Abe M, Al-Mahtab M, Furukawa S, Bunzo M, Hiasa Y, Onji M (2015) Diminished immune response to vaccinations in obesity: role of myeloid-derived suppressor and other myeloid cells. Obes Res Clin Pract 9(1):35–44. doi:10.1016/j.orcp.2013.12.006

39. Rodriguez PC, Ernstoff MS, Hernandez C, Atkins M, Zabaleta J, Sierra R, Ochoa AC (2009) Arginase I-producing myeloid-derived suppressor cells in renal cell carcinoma are a subpopulation of activated granulocytes. Cancer Res 69(4):1553–1560. doi:10.1158/0008-5472.CAN-08-1921

40. Lin Y, Gustafson MP, Bulur PA, Gastineau DA, Witzig TE, Dietz AB (2011) Immunosuppressive CD14+ HLA-DR(low)/- monocytes in B-cell non-Hodgkin lymphoma. Blood 117(3):872–881. doi:10.1182/blood-2010-05-283820

41. Solito S, Bronte V, Mandruzzato S (2011) Antigen specificity of immune suppression by myeloid-derived suppressor cells. J Leukoc Biol 90(1):31–36. doi:10.1189/jlb.0111021

42. Gabrilovich DI, Velders MP, Sotomayor EM, Kast WM (2001) Mechanism of immune dysfunction in cancer mediated by immature Gr-1+ myeloid cells. J Immunol 166(9):5398–5406

43. Kusmartsev S, Nagaraj S, Gabrilovich DI (2005) Tumor-associated CD8+ T cell tolerance induced by bone marrow-derived immature myeloid cells. J Immunol 175(7):4583–4592

44. Kusmartsev S, Nefedova Y, Yoder D, Gabrilovich DI (2004) Antigen-specific inhibition of CD8+ T cell response by immature myeloid cells in cancer is mediated by reactive oxygen species. J Immunol 172(2):989–999

45. Serafini P, Mgebroff S, Noonan K, Borrello I (2008) Myeloid-derived suppressor cells promote cross-tolerance in B-cell lymphoma by expanding regulatory T cells. Cancer Res 68 (13):5439–5449. doi:10.1158/0008-5472.CAN-07-6621

46. Chalmin F, Ladoire S, Mignot G, Vincent J, Bruchard M, Remy-Martin JP, Boireau W, Rouleau A, Simon B, Lanneau D, De Thonel A, Multhoff G, Hamman A, Martin F, Chauffert B, Solary E, Zitvogel L, Garrido C, Ryffel B, Borg C, Apetoh L, Rebe C, Ghiringhelli F (2010) Membrane-associated Hsp72 from tumor-derived exosomes mediates STAT3-dependent immunosuppressive function of mouse and human myeloid-derived suppressor cells. J Clin Investig 120(2):457–471. doi:10.1172/JCI40483

47. Kodumudi KN, Woan K, Gilvary DL, Sahakian E, Wei S, Djeu JY (2010) A novel chemoimmunomodulating property of docetaxel: suppression of myeloid-derived suppressor cells in tumor bearers. Clin Cancer Res Official J Am Assoc Cancer Res 16(18):4583–4594. doi:10.1158/1078-0432.CCR-10-0733

48. Sinha P, Clements VK, Ostrand-Rosenberg S (2005) Reduction of myeloid-derived suppressor cells and induction of M1 macrophages facilitate the rejection of established metastatic disease. J Immunol 174(2):636–645

49. Nagaraj S, Nelson A, Youn JI, Cheng P, Quiceno D, Gabrilovich DI (2012) Antigen-specific CD4(+) T cells regulate function of myeloid-derived suppressor cells in cancer via retrograde MHC class II signaling. Cancer Res 72(4):928–938. doi:10.1158/0008-5472.CAN-11-2863

50. Fridlender ZG, Sun J, Kim S, Kapoor V, Cheng G, Ling L, Worthen GS, Albelda SM (2009) Polarization of tumor-associated neutrophil phenotype by TGF-beta: "N1" versus "N2" TAN. Cancer Cell 16(3):183–194. doi:10.1016/j.ccr.2009.06.017

51. Mishalian I, Bayuh R, Levy L, Zolotarov L, Michaeli J, Fridlender ZG (2013) Tumor-associated neutrophils (TAN) develop pro-tumorigenic properties during tumor progression. Cancer Immunol Immunother CII 62(11):1745–1756. doi:10.1007/s00262-013-1476-9

52. Youn JI, Collazo M, Shalova IN, Biswas SK, Gabrilovich DI (2012) Characterization of the nature of granulocytic myeloid-derived suppressor cells in tumor-bearing mice. J Leukoc Biol 91(1):167–181. doi:10.1189/jlb.0311177

53. Schmielau J, Finn OJ (2001) Activated granulocytes and granulocyte-derived hydrogen peroxide are the underlying mechanism of suppression of t-cell function in advanced cancer patients. Cancer Res 61(12):4756–4760

54. Trellakis S, Bruderek K, Hutte J, Elian M, Hoffmann TK, Lang S, Brandau S (2013) Granulocytic myeloid-derived suppressor cells are cryosensitive and their frequency does not correlate with serum concentrations of colony-stimulating factors in head and neck cancer. Innate Immunity 19(3):328–336. doi:10.1177/1753425912463618

55. Elghetany MT (2002) Surface antigen changes during normal neutrophilic development: a critical review. Blood Cells Mol Dis 28(2):260–274

56. Choi J, Suh B, Ahn YO, Kim TM, Lee JO, Lee SH, Heo DS (2012) CD15+/CD16low human granulocytes from terminal cancer patients: granulocytic myeloid-derived suppressor cells that have suppressive function. Tumour Biol J Int Soc Oncodevelopmental Biol Med 33 (1):121–129. doi:10.1007/s13277-011-0254-6

57. Martinez FO, Helming L, Gordon S (2009) Alternative activation of macrophages: an immunologic functional perspective. Ann Rev Immunol 27:451–483. doi:10.1146/annurev.immunol.021908.132532

58. de Goeje PL, Bezemer K, Heuvers ME, Dingemans AC, Groen HJ, Smit EF, Hoogsteden HC, Hendriks RW, Aerts JG, Hegmans JP (2015) Immunoglobulin-like transcript 3 is expressed by myeloid-derived suppressor cells and correlates with survival in patients with non-small cell lung cancer. Oncoimmunology 4(7):e1014242. doi:10.1080/2162402X.2015.1014242

59. Solito S, Marigo I, Pinton L, Damuzzo V, Mandruzzato S, Bronte V (2014) Myeloid-derived suppressor cell heterogeneity in human cancers. Ann NY Acad Sci 1319:47–65. doi:10.1111/nyas.12469

60. Abe F, Dafferner AJ, Donkor M, Westphal SN, Scholar EM, Solheim JC, Singh RK, Hoke TA, Talmadge JE (2010) Myeloid-derived suppressor cells in mammary tumor progression in FVB Neu transgenic mice. Cancer Immunol Immunother CII 59(1):47–62. doi:10.1007/s00262-009-0719-2

61. Lu P, Yu B, Xu J (2012) Cucurbitacin B regulates immature myeloid cell differentiation and enhances antitumor immunity in patients with lung cancer. Cancer Biother Radiopharm 27 (8):495–503. doi:10.1089/cbr.2012.1219

62. Serafini P, Meckel K, Kelso M, Noonan K, Califano J, Koch W, Dolcetti L, Bronte V, Borrello I (2006) Phosphodiesterase-5 inhibition augments endogenous antitumor immunity by reducing myeloid-derived suppressor cell function. J Exp Med 203(12):2691–2702. doi:10.1084/jem.20061104

63. Califano JA, Khan Z, Noonan KA, Rudraraju L, Zhang Z, Wang H, Goodman S, Gourin CG, Ha PK, Fakhry C, Saunders J, Levine M, Tang M, Neuner G, Richmon JD, Blanco R, Agrawal N, Koch WM, Marur S, Weed DT, Serafini P, Borrello I (2015) Tadalafil augments tumor specific immunity in patients with head and neck squamous cell carcinoma. Clin Cancer Res Official J Am Assoc Cancer Res 21(1):30–38. doi:10.1158/1078-0432.CCR-14-1716

64. Noonan KA, Ghosh N, Rudraraju L, Bui M, Borrello I (2014) Targeting immune suppression with PDE5 inhibition in end-stage multiple myeloma. Cancer Immunol Res 2(8):725–731. doi:10.1158/2326-6066.CIR-13-0213

65. Weed DT, Vella JL, Reis IM, De la Fuente AC, Gomez C, Sargi Z, Nazarian R, Califano J, Borrello I, Serafini P (2015) Tadalafil reduces myeloid-derived suppressor cells and regulatory T cells and promotes tumor immunity in patients with head and neck squamous cell carcinoma. Clin Cancer Res Official J Am Assoc Cancer Res 21(1):39–48. doi:10.1158/1078-0432.CCR-14-1711

66. Roth F, De La Fuente AC, Vella JL, Zoso A, Inverardi L, Serafini P (2012) Aptamer-mediated blockade of IL4Ralpha triggers apoptosis of MDSCs and limits tumor progression. Cancer Res 72(6):1373–1383. doi:10.1158/0008-5472.CAN-11-2772

67. Sinha P, Parker KH, Horn L, Ostrand-Rosenberg S (2012) Tumor-induced myeloid-derived suppressor cell function is independent of IFN-gamma and IL-4Ralpha. Eur J Immunol 42(8):2052–2059. doi:10.1002/eji.201142230

68. De Santo C, Serafini P, Marigo I, Dolcetti L, Bolla M, Del Soldato P, Melani C, Guiducci C, Colombo MP, Iezzi M, Musiani P, Zanovello P, Bronte V (2005) Nitroaspirin corrects immune dysfunction in tumor-bearing hosts and promotes tumor eradication by cancer vaccination. Proc Natl Acad Sci USA 102(11):4185–4190. doi:10.1073/pnas.0409783102

69. Thimmulappa RK, Fuchs RJ, Malhotra D, Scollick C, Traore K, Bream JH, Trush MA, Liby KT, Sporn MB, Kensler TW, Biswal S (2007) Preclinical evaluation of targeting the Nrf2 pathway by triterpenoids (CDDO-Im and CDDO-Me) for protection from LPS-induced inflammatory response and reactive oxygen species in human peripheral blood mononuclear cells and neutrophils. Antioxid Redox Sig 9(11):1963–1970. doi:10.1089/ars.2007.1745

70. Nagaraj S, Youn JI, Weber H, Iclozan C, Lu L, Cotter MJ, Meyer C, Becerra CR, Fishman M, Antonia S, Sporn MB, Liby KT, Rawal B, Lee JH, Gabrilovich DI (2010) Anti-inflammatory triterpenoid blocks immune suppressive function of MDSCs and improves immune response in cancer. Clin Cancer Res Official J Am Assoc Cancer Res 16(6):1812–1823. doi:10.1158/1078-0432.CCR-09-3272

71. Zha S, Yegnasubramanian V, Nelson WG, Isaacs WB, De Marzo AM (2004) Cyclooxygenases in cancer: progress and perspective. Cancer Lett 215(1):1–20. doi:10.1016/j.canlet.2004.06.014

72. Rodriguez PC, Hernandez CP, Quiceno D, Dubinett SM, Zabaleta J, Ochoa JB, Gilbert J, Ochoa AC (2005) Arginase I in myeloid suppressor cells is induced by COX-2 in lung carcinoma. J Exp Med 202(7):931–939. doi:10.1084/jem.20050715

73. Veltman JD, Lambers ME, van Nimwegen M, Hendriks RW, Hoogsteden HC, Aerts JG, Hegmans JP (2010) COX-2 inhibition improves immunotherapy and is associated with decreased numbers of myeloid-derived suppressor cells in mesothelioma. Celecoxib influences MDSC function. BMC Cancer 10:464. doi:10.1186/1471-2407-10-464

74. Bracci L, Schiavoni G, Sistigu A, Belardelli F (2014) Immune-based mechanisms of cytotoxic chemotherapy: implications for the design of novel and rationale-based combined treatments against cancer. Cell Death Differ 21(1):15–25. doi:10.1038/cdd.2013.67

75. Suzuki E, Kapoor V, Jassar AS, Kaiser LR, Albelda SM (2005) Gemcitabine selectively eliminates splenic Gr-1$^+$/CD11b$^+$ myeloid suppressor cells in tumor-bearing animals and enhances antitumor immune activity. Clin Cancer Res Official J Am Assoc Cancer Res 11(18):6713–6721. doi:10.1158/1078-0432.CCR-05-0883

76. Vincent J, Mignot G, Chalmin F, Ladoire S, Bruchard M, Chevriaux A, Martin F, Apetoh L, Rebe C, Ghiringhelli F (2010) 5-Fluorouracil selectively kills tumor-associated myeloid-derived suppressor cells resulting in enhanced T cell-dependent antitumor immunity. Cancer Res 70(8):3052–3061. doi:10.1158/0008-5472.CAN-09-3690

77. Chen J, Huang X, Huang G, Chen Y, Chen L, Song H (2012) Preconditioning chemotherapy with cisplatin enhances the antitumor activity of cytokine-induced killer cells in a murine melanoma model. Cancer Biother Radiopharm 27(3):210–220. doi:10.1089/cbr.2011.1116

78. Naiditch H, Shurin MR, Shurin GV (2011) Targeting myeloid regulatory cells in cancer by chemotherapeutic agents. Immunol Res 50(2–3):276–285. doi:10.1007/s12026-011-8213-2

79. Diaz-Montero CM, Salem ML, Nishimura MI, Garrett-Mayer E, Cole DJ, Montero AJ (2009) Increased circulating myeloid-derived suppressor cells correlate with clinical cancer stage, metastatic tumor burden, and doxorubicin-cyclophosphamide chemotherapy. Cancer Immunol Immunother CII 58(1):49–59. doi:10.1007/s00262-008-0523-4

80. Tongu M, Harashima N, Monma H, Inao T, Yamada T, Kawauchi H, Harada M (2013) Metronomic chemotherapy with low-dose cyclophosphamide plus gemcitabine can induce anti-tumor T cell immunity in vivo. Cancer Immunol Immunother CII 62(2):383–391. doi:10.1007/s00262-012-1343-0

81. Yanagimoto H, Mine T, Yamamoto K, Satoi S, Terakawa N, Takahashi K, Nakahara K, Honma S, Tanaka M, Mizoguchi J, Yamada A, Oka M, Kamiyama Y, Itoh K, Takai S (2007) Immunological evaluation of personalized peptide vaccination with gemcitabine for pancreatic cancer. Cancer Sci 98(4):605–611. doi:10.1111/j.1349-7006.2007.00429.x

82. Annels NE, Shaw VE, Gabitass RF, Billingham L, Corrie P, Eatock M, Valle J, Smith D, Wadsley J, Cunningham D, Pandha H, Neoptolemos JP, Middleton G (2014) The effects of gemcitabine and capecitabine combination chemotherapy and of low-dose adjuvant GM-CSF on the levels of myeloid-derived suppressor cells in patients with advanced pancreatic cancer. Cancer Immunol Immunother CII 63(2):175–183. doi:10.1007/s00262-013-1502-y

83. Montero AJ, Diaz-Montero CM, Deutsch YE, Hurley J, Koniaris LG, Rumboldt T, Yasir S, Jorda M, Garret-Mayer E, Avisar E, Slingerland J, Silva O, Welsh C, Schuhwerk K, Seo P, Pegram MD, Gluck S (2012) Phase 2 study of neoadjuvant treatment with NOV-002 in combination with doxorubicin and cyclophosphamide followed by docetaxel in patients with HER-2 negative clinical stage II-IIIc breast cancer. Breast Cancer Res Treat 132(1):215–223. doi:10.1007/s10549-011-1889-0

84. Drake MT, Clarke BL, Khosla S (2008) Bisphosphonates: mechanism of action and role in clinical practice. Mayo Clin Proc 83(9):1032–1045. doi:10.4065/83.9.1032

85. Melani C, Sangaletti S, Barazzetta FM, Werb Z, Colombo MP (2007) Amino-biphosphonate-mediated MMP-9 inhibition breaks the tumor-bone marrow axis responsible for myeloid-derived suppressor cell expansion and macrophage infiltration in tumor stroma. Cancer Res 67(23):11438–11446. doi:10.1158/0008-5472.CAN-07-1882

86. Porembka MR, Mitchem JB, Belt BA, Hsieh CS, Lee HM, Herndon J, Gillanders WE, Linehan DC, Goedegebuure P (2012) Pancreatic adenocarcinoma induces bone marrow mobilization of myeloid-derived suppressor cells which promote primary tumor growth. Cancer Immunol Immunother CII 61(9):1373–1385. doi:10.1007/s00262-011-1178-0

87. Sanford DE, Porembka MR, Panni RZ, Mitchem JB, Belt BA, Plambeck-Suess SM, Lin G, Denardo DG, Fields RC, Hawkins WG, Strasberg SM, Lockhart AC, Wang-Gillam A, Goedegebuure SP, Linehan DC (2013) A study of zoledronic acid as neo-adjuvant, perioperative therapy in patients with resectable pancreatic ductal adenocarcinoma. J Cancer Ther 4(3):797–803. doi:10.4236/jct.2013.43096

88. Gnant M, Mlineritsch B, Stoeger H, Luschin-Ebengreuth G, Heck D, Menzel C, Jakesz R, Seifert M, Hubalek M, Pristauz G, Bauernhofer T, Eidtmann H, Eiermann W, Steger G, Kwasny W, Dubsky P, Hochreiner G, Forsthuber EP, Fesl C, Greil R, Austrian B, Colorectal Cancer Study Group VA (2011) Adjuvant endocrine therapy plus zoledronic acid in premenopausal women with early-stage breast cancer: 62-month follow-up from the ABCSG-12 randomised trial. Lancet Oncol 12 (7):631–641. doi:10.1016/S1470-2045(11)70122-X

89. Eidtmann H, de Boer R, Bundred N, Llombart-Cussac A, Davidson N, Neven P, von Minckwitz G, Miller J, Schenk N, Coleman R (2010) Efficacy of zoledronic acid in postmenopausal women with early breast cancer receiving adjuvant letrozole: 36-month results of the ZO-FAST study. Ann Oncol Official J Eur Soc Med Oncol ESMO 21 (11):2188–2194. doi:10.1093/annonc/mdq217

90. Coleman RMG, Paterson A, Powles T, von Minckwitz G, Pritchard K, Bergh J, Bliss J, Gralow J, Anderson S, Evans V, Pan H, Bradley R, Davies C, and Gray R (2013) Abstract S4-07: Effects of bisphosphonate treatment on recurrence and cause-specific mortality in women with early breast cancer: a meta-analysis of individual patient data from randomised trials. Cancer Res 73

91. Coleman RE, Marshall H, Cameron D, Dodwell D, Burkinshaw R, Keane M, Gil M, Houston SJ, Grieve RJ, Barrett-Lee PJ, Ritchie D, Pugh J, Gaunt C, Rea U, Peterson J, Davies C, Hiley V, Gregory W, Bell R, Investigators A (2011) Breast-cancer adjuvant therapy with zoledronic acid. N Engl J Med 365(15):1396–1405. doi:10.1056/NEJMoa1105195

92. Nienhuis HH, Gaykema SB, Timmer-Bosscha H, Jalving M, Brouwers AH, Lub-de Hooge MN, van der Vegt B, Overmoyer B, de Vries EG, Schroder CP (2015) Targeting breast cancer through its microenvironment: current status of preclinical and clinical research in finding relevant targets. Pharmacol Ther 147:63–79. doi:10.1016/j.pharmthera.2014.11.004

93. Cheng P, Corzo CA, Luetteke N, Yu B, Nagaraj S, Bui MM, Ortiz M, Nacken W, Sorg C, Vogl T, Roth J, Gabrilovich DI (2008) Inhibition of dendritic cell differentiation and accumulation of myeloid-derived suppressor cells in cancer is regulated by S100A9 protein. J Exp Med 205(10):2235–2249. doi:10.1084/jem.20080132

94. Sinha P, Okoro C, Foell D, Freeze HH, Ostrand-Rosenberg S, Srikrishna G (2008) Proinflammatory S100 proteins regulate the accumulation of myeloid-derived suppressor cells. J Immunol 181(7):4666–4675

95. Xu J, Escamilla J, Mok S, David J, Priceman S, West B, Bollag G, McBride W, Wu L (2013) CSF1R signaling blockade stanches tumor-infiltrating myeloid cells and improves the efficacy of radiotherapy in prostate cancer. Cancer Res 73(9):2782–2794. doi:10.1158/0008-5472.CAN-12-3981

96. Waight JD, Hu Q, Miller A, Liu S, Abrams SI (2011) Tumor-derived G-CSF facilitates neoplastic growth through a granulocytic myeloid-derived suppressor cell-dependent mechanism. PLoS ONE 6(11):e27690. doi:10.1371/journal.pone.0027690

97. Bayne LJ, Beatty GL, Jhala N, Clark CE, Rhim AD, Stanger BZ, Vonderheide RH (2012) Tumor-derived granulocyte-macrophage colony-stimulating factor regulates myeloid inflammation and T cell immunity in pancreatic cancer. Cancer Cell 21(6):822–835. doi:10.1016/j.ccr.2012.04.025

98. Pan PY, Wang GX, Yin B, Ozao J, Ku T, Divino CM, Chen SH (2008) Reversion of immune tolerance in advanced malignancy: modulation of myeloid-derived suppressor cell development by blockade of stem-cell factor function. Blood 111(1):219–228. doi:10.1182/blood-2007-04-086835

99. Serafini P, Carbley R, Noonan KA, Tan G, Bronte V, Borrello I (2004) High-dose granulocyte-macrophage colony-stimulating factor-producing vaccines impair the immune response through the recruitment of myeloid suppressor cells. Cancer Res 64(17):6337–6343. doi:10.1158/0008-5472.CAN-04-0757

100. Morales JK, Kmieciak M, Knutson KL, Bear HD, Manjili MH (2010) GM-CSF is one of the main breast tumor-derived soluble factors involved in the differentiation of CD11b-Gr1- bone marrow progenitor cells into myeloid-derived suppressor cells. Breast Cancer Res Treat 123(1):39–49. doi:10.1007/s10549-009-0622-8

101. Filipazzi P, Valenti R, Huber V, Pilla L, Canese P, Iero M, Castelli C, Mariani L, Parmiani G, Rivoltini L (2007) Identification of a new subset of myeloid suppressor cells in peripheral blood of melanoma patients with modulation by a granulocyte-macrophage colony-stimulation factor-based antitumor vaccine. J Clin Oncol Official J Am Soc Clin Oncol 25(18):2546–2553. doi:10.1200/JCO.2006.08.5829

102. Terme M, Colussi O, Marcheteau E, Tanchot C, Tartour E, Taieb J (2012) Modulation of immunity by antiangiogenic molecules in cancer. Clin Dev Immunol 2012:492920. doi:10.1155/2012/492920

103. Ko JS, Zea AH, Rini BI, Ireland JL, Elson P, Cohen P, Golshayan A, Rayman PA, Wood L, Garcia J, Dreicer R, Bukowski R, Finke JH (2009) Sunitinib mediates reversal of myeloid-derived suppressor cell accumulation in renal cell carcinoma patients. Clin Cancer Res Official J Am Assoc Cancer Res 15(6):2148–2157. doi:10.1158/1078-0432.CCR-08-1332

104. Fricke I, Mirza N, Dupont J, Lockhart C, Jackson A, Lee JH, Sosman JA, Gabrilovich DI (2007) Vascular endothelial growth factor-trap overcomes defects in dendritic cell differentiation but does not improve antigen-specific immune responses. Clin Cancer Res Official J Am Assoc Cancer Res 13(16):4840–4848. doi:10.1158/1078-0432.CCR-07-0409

105. Osada T, Chong G, Tansik R, Hong T, Spector N, Kumar R, Hurwitz HI, Dev I, Nixon AB, Lyerly HK, Clay T, Morse MA (2008) The effect of anti-VEGF therapy on immature myeloid cell and dendritic cells in cancer patients. Cancer Immunol Immunother CII 57 (8):1115–1124. doi:10.1007/s00262-007-0441-x

106. Schilling B, Sucker A, Griewank K, Zhao F, Weide B, Gorgens A, Giebel B, Schadendorf D, Paschen A (2013) Vemurafenib reverses immunosuppression by myeloid derived suppressor cells. Int J Cancer 133(7):1653–1663. doi:10.1002/ijc.28168

107. Najjar YG, Finke JH (2013) Clinical perspectives on targeting of myeloid derived suppressor cells in the treatment of cancer. Front Oncol 3:49. doi:10.3389/fonc.2013.00049

108. Kusmartsev S, Cheng F, Yu B, Nefedova Y, Sotomayor E, Lush R, Gabrilovich D (2003) All-trans-retinoic acid eliminates immature myeloid cells from tumor-bearing mice and improves the effect of vaccination. Cancer Res 63(15):4441–4449

109. Kusmartsev S, Su Z, Heiser A, Dannull J, Eruslanov E, Kubler H, Yancey D, Dahm P, Vieweg J (2008) Reversal of myeloid cell-mediated immunosuppression in patients with metastatic renal cell carcinoma. Clin Cancer Res Official J Am Assoc Cancer Res 14 (24):8270–8278. doi:10.1158/1078-0432.CCR-08-0165

110. Iclozan C, Antonia S, Chiappori A, Chen DT, Gabrilovich D (2013) Therapeutic regulation of myeloid-derived suppressor cells and immune response to cancer vaccine in patients with extensive stage small cell lung cancer. Cancer Immunol Immunother CII 62(5):909–918. doi:10.1007/s00262-013-1396-8

111. Kulbersh JS, Day TA, Gillespie MB, Young MR (2009) 1alpha,25-Dihydroxyvitamin D(3) to skew intratumoral levels of immune inhibitory CD34(+) progenitor cells into dendritic cells. Otolaryngol Head Neck Surg Official J Am Acad Otolaryngol Head Neck Surg 140 (2):235–240. doi:10.1016/j.otohns.2008.11.011

112. Obermajer N, Muthuswamy R, Lesnock J, Edwards RP, Kalinski P (2011) Positive feedback between PGE2 and COX2 redirects the differentiation of human dendritic cells toward stable myeloid-derived suppressor cells. Blood 118(20):5498–5505. doi:10.1182/blood-2011-07-365825

113. Mirza N, Fishman M, Fricke I, Dunn M, Neuger AM, Frost TJ, Lush RM, Antonia S, Gabrilovich DI (2006) All-trans-retinoic acid improves differentiation of myeloid cells and immune response in cancer patients. Cancer Res 66(18):9299–9307. doi:10.1158/0008-5472.CAN-06-1690

114. Lathers DM, Clark JI, Achille NJ, Young MR (2004) Phase 1B study to improve immune responses in head and neck cancer patients using escalating doses of 25-hydroxyvitamin D3. Cancer Immunol Immunother CII 53(5):422–430. doi:10.1007/s00262-003-0459-7

115. Young MR, Lathers DM (1999) Myeloid progenitor cells mediate immune suppression in patients with head and neck cancers. Int J Immunopharmacol 21(4):241–252

116. Young MR, Ihm J, Lozano Y, Wright MA, Prechel MM (1995) Treating tumor-bearing mice with vitamin D3 diminishes tumor-induced myelopoiesis and associated immunosuppression, and reduces tumor metastasis and recurrence. Cancer Immunol Immunother CII 41(1):37–45

117. Umansky V, Sevko A (2012) Overcoming immunosuppression in the melanoma microenvironment induced by chronic inflammation. Cancer Immunol Immunother CII 61 (2):275–282. doi:10.1007/s00262-011-1164-6

118. Pratima Sinha VKC, Stephanie K. Bunt, Steven M. Albelda and Suzanne Ostrand-Rosenberg (2007) Cross-talk between myeloid-derived suppressor cells and macrophages subverts tumor immunity toward a type 2 response. J Immunol 179(2):977–983

119. Kusmartsev S, Eruslanov E, Kubler H, Tseng T, Sakai Y, Su Z, Kaliberov S, Heiser A, Rosser C, Dahm P, Siemann D, Vieweg J (2008) Oxidative stress regulates expression of VEGFR1 in myeloid cells: link to tumor-induced immune suppression in renal cell carcinoma. J Immunol 181(1):346–353

120. Yang L, Huang J, Ren X, Gorska AE, Chytil A, Aakre M, Carbone DP, Matrisian LM, Richmond A, Lin PC, Moses HL (2008) Abrogation of TGF beta signaling in mammary carcinomas recruits Gr-1$^+$ CD11b$^+$ myeloid cells that promote metastasis. Cancer Cell 13 (1):23–35. doi:10.1016/j.ccr.2007.12.004

121. van Cruijsen H, van der Veldt AA, Vroling L, Oosterhoff D, Broxterman HJ, Scheper RJ, Giaccone G, Haanen JB, van den Eertwegh AJ, Boven E, Hoekman K, de Gruijl TD (2008) Sunitinib-induced myeloid lineage redistribution in renal cell cancer patients: CD1c$^+$ dendritic cell frequency predicts progression-free survival. Clin Cancer Res Official J Am Assoc Cancer Res 14(18):5884–5892. doi:10.1158/1078-0432.CCR-08-0656

122. Shojaei F, Wu X, Zhong C, Yu L, Liang XH, Yao J, Blanchard D, Bais C, Peale FV, van Bruggen N, Ho C, Ross J, Tan M, Carano RA, Meng YG, Ferrara N (2007) Bv8 regulates myeloid-cell-dependent tumour angiogenesis. Nature 450(7171):825–831. doi:10.1038/nature06348

123. Collazo MM, Paraiso KH, Park MY, Hazen AL, Kerr WG (2012) Lineage extrinsic and intrinsic control of immunoregulatory cell numbers by SHIP. Eur J Immunol 42(7):1785–1795. doi:10.1002/eji.201142092

124. Shojaei F, Wu X, Qu X, Kowanetz M, Yu L, Tan M, Meng YG, Ferrara N (2009) G-CSF-initiated myeloid cell mobilization and angiogenesis mediate tumor refractoriness to anti-VEGF therapy in mouse models. Proc Natl Acad Sci USA 106(16):6742–6747. doi:10.1073/pnas.0902280106

125. DeNardo DG, Brennan DJ, Rexhepaj E, Ruffell B, Shiao SL, Madden SF, Gallagher WM, Wadhwani N, Keil SD, Junaid SA, Rugo HS, Hwang ES, Jirstrom K, West BL, Coussens LM (2011) Leukocyte complexity predicts breast cancer survival and functionally regulates response to chemotherapy. Cancer Discov 1(1):54–67. doi:10.1158/2159-8274.CD-10-0028

126. Fernandez A, Mesa C, Marigo I, Dolcetti L, Clavell M, Oliver L, Fernandez LE, Bronte V (2011) Inhibition of tumor-induced myeloid-derived suppressor cell function by a nanoparticulated adjuvant. J Immunol 186(1):264–274. doi:10.4049/jimmunol.1001465

127. Loberg RD, Ying C, Craig M, Day LL, Sargent E, Neeley C, Wojno K, Snyder LA, Yan L, Pienta KJ (2007) Targeting CCL2 with systemic delivery of neutralizing antibodies induces prostate cancer tumor regression in vivo. Cancer Res 67(19):9417–9424. doi:10.1158/0008-5472.CAN-07-1286

128. Kerkar SP, Goldszmid RS, Muranski P, Chinnasamy D, Yu Z, Reger RN, Leonardi AJ, Morgan RA, Wang E, Marincola FM, Trinchieri G, Rosenberg SA, Restifo NP (2011) IL-12 triggers a programmatic change in dysfunctional myeloid-derived cells within mouse tumors. J Clin Investig 121(12):4746–4757. doi:10.1172/JCI58814

129. Chmielewski M, Kopecky C, Hombach AA, Abken H (2011) IL-12 release by engineered T cells expressing chimeric antigen receptors can effectively muster an antigen-independent macrophage response on tumor cells that have shut down tumor antigen expression. Cancer Res 71(17):5697–5706. doi:10.1158/0008-5472.CAN-11-0103

130. Weiss JM, Ridnour LA, Back T, Hussain SP, He P, Maciag AE, Keefer LK, Murphy WJ, Harris CC, Wink DA, Wiltrout RH (2010) Macrophage-dependent nitric oxide expression regulates tumor cell detachment and metastasis after IL-2/anti-CD40 immunotherapy. J Exp Med 207(11):2455–2467. doi:10.1084/jem.20100670

131. Beatty GL, Chiorean EG, Fishman MP, Saboury B, Teitelbaum UR, Sun W, Huhn RD, Song W, Li D, Sharp LL, Torigian DA, O'Dwyer PJ, Vonderheide RH (2011) CD40 agonists alter tumor stroma and show efficacy against pancreatic carcinoma in mice and humans. Science 331(6024):1612–1616. doi:10.1126/science.1198443

132. Rolny C, Mazzone M, Tugues S, Laoui D, Johansson I, Coulon C, Squadrito ML, Segura I, Li X, Knevels E, Costa S, Vinckier S, Dresselaer T, Akerud P, De Mol M, Salomaki H, Phillipson M, Wyns S, Larsson E, Buysschaert I, Botling J, Himmelreich U, Van Ginderachter JA, De Palma M, Dewerchin M, Claesson-Welsh L, Carmeliet P (2011) HRG inhibits tumor growth and metastasis by inducing macrophage polarization and vessel normalization through downregulation of PlGF. Cancer Cell 19(1):31–44. doi:10.1016/j.ccr.2010.11.009

133. Hagemann T, Lawrence T, McNeish I, Charles KA, Kulbe H, Thompson RG, Robinson SC, Balkwill FR (2008) "Re-educating" tumor-associated macrophages by targeting NF-kappaB. J Exp Med 205(6):1261–1268. doi:10.1084/jem.20080108

134. Davies GF, Khandelwal RL, Wu L, Juurlink BH, Roesler WJ (2001) Inhibition of phosphoenolpyruvate carboxykinase (PEPCK) gene expression by troglitazone: a peroxisome proliferator-activated receptor-gamma (PPARgamma)-independent, antioxidant-related mechanism. Biochem Pharmacol 62(8):1071–1079

135. Sumida K, Wakita D, Narita Y, Masuko K, Terada S, Watanabe K, Satoh T, Kitamura H, Nishimura T (2012) Anti-IL-6 receptor mAb eliminates myeloid-derived suppressor cells and inhibits tumor growth by enhancing T-cell responses. Eur J Immunol 42(8):2060–2072. doi:10.1002/eji.201142335

136. Ries CH, Cannarile MA, Hoves S, Benz J, Wartha K, Runza V, Rey-Giraud F, Pradel LP, Feuerhake F, Klaman I, Jones T, Jucknischke U, Scheiblich S, Kaluza K, Gorr IH, Walz A, Abiraj K, Cassier PA, Sica A, Gomez-Roca C, de Visser KE, Italiano A, Le Tourneau C, Delord JP, Levitsky H, Blay JY, Ruttinger D (2014) Targeting tumor-associated macrophages with anti-CSF-1R antibody reveals a strategy for cancer therapy. Cancer Cell 25(6):846–859. doi:10.1016/j.ccr.2014.05.016

137. Meyer C, Cagnon L, Costa-Nunes CM, Baumgaertner P, Montandon N, Leyvraz L, Michielin O, Romano E, Speiser DE (2014) Frequencies of circulating MDSC correlate with clinical outcome of melanoma patients treated with ipilimumab. Cancer Immunol Immunother CII 63(3):247–257. doi:10.1007/s00262-013-1508-5

138. Gorgun G, Samur MK, Cowens KB, Paula S, Bianchi G, Anderson JE, White RE, Singh A, Ohguchi H, Suzuki R, Kikuchi S, Harada T, Hideshima T, Tai YT, Laubach JP, Raje N, Magrangeas F, Minvielle S, Avet-Loiseau H, Munshi NC, Dorfman DM, Richardson PG, Anderson KC (2015) Lenalidomide enhances immune checkpoint blockade-induced immune response in multiple myeloma. Clin Cancer Res Official J Am Assoc Cancer Res 21 (20):4607–4618. doi:10.1158/1078-0432.CCR-15-0200

Chapter 2
Differentiation of Murine Myeloid-Derived Suppressor Cells

David Escors

Abstract Myeloid-derived suppressor cells (MDSCs) are frequently defined as a heterogeneous population of immature cells belonging to the myeloid lineage which possess strong immunosuppressive activities. These cells ultimately derive from myeloid progenitors mainly present in the bone marrow that undergo a dysregulated differentiation pathway, ending up with the systemic mobilization of MDSCs of "monocytic" or "granulocytic" characteristics. Here we will review the current knowledge on MDSC differentiation mostly in murine cancer models, and reflect on whether MDSCs represent a unique, well-defined distinct myeloid lineage or just immature stages of myeloid cells.

Keywords Haematopoiesis · Myeloid lineage · Granulocytes · Monocytes · Dendritic cells · Myeloid-derived suppressor cells · Bone marrow

2.1 Introduction

An extensive review of the specialized literature quickly shows to the non-specialized reader that myeloid-derived suppressor cells (MDSCs) are defined in many ways, depending on the cancer model and the particular standards adopted by each research team. These variable definitions widely used by researchers highlight the ultimate question in MDSC research: Do MDSCs belong to a distinct, genuine myeloid lineage or are they a collection of myeloid cells halted at several differentiation steps of a disturbed myelopoiesis?

MDSCs are most commonly defined as a "heterogenous population of immature cells of myeloid origin with strong suppressive activities". It is also assumed by

D. Escors (✉)
Navarrabiomed-Biomedical Research Centre, Fundación Miguel Servet, IdiSNA,
Calle Irunlarrea 3, Pamplona, Navarra 31008, Spain
e-mail: descorsm@navarra.es

D. Escors
Rayne Institute. University College London, 5 University Street, London WC1E 6JF, UK

© The Author(s) 2016
D. Escors et al., *Myeloid-Derived Suppressor Cells and Cancer*,
SpringerBriefs in Immunology, DOI 10.1007/978-3-319-26821-7_2

many researchers that MDSCs arise from two independent myeloid lineages, namely, monocytic and granulocytic lineages. A significant number of researchers define these cells as "dysfunctional inflammatory monocytes" (for monocytic MDSCs) or "dysfunctional neutrophils" (for granulocytic MDSCs). Finally, others have adopted a nomenclature that englobes all possible immunosuppressive cells of myeloid origin without any attempt at their distinction or classification; the "immature suppressive myeloid cells" or "myeloid regulatory cells".

Nevertheless, the existence and nature of MDSCs cannot be ignored, and the ultimate question on the origin of MDSCs must be consequently addressed. First, it will expand our knowledge on how cancer subverts our immune system to its advantage. Second, from a therapeutic point of view, it is necessary to find out if we are working with a single differentiation pathway, or if immunosuppressive MDSCs arise independently from multiple precursors by redundant differentiation pathways.

In this chapter we will revise the physiological myeloid differentiation pathway, MDSC differentiation, and whether we can consider MDSCs as a genuine myeloid lineage arising from a well-defined differentiation pathway.

2.2   Physiological Myeloid Differentiation

2.2.1   Conventional Myeloid Cell Types and Their Role in Immunity

Myeloid cells play key regulatory and effector roles in both innate and adaptive immunity. Some of the myeloid cell lineages possess very strong antigen-presenting capacities, and they express a wide range of pathogen pattern-recognition receptors on their surface. These receptors, when engaged by their respective ligands, start a process of "maturation" in these cells leading to characteristic phenotypic and functional changes resulting in cytokine secretion, production of a wide range of biologically active molecules and acquisition of strong antigen-presenting capacities. Classically, "mature" myeloid cells englobe four main well-defined classical cell types. Nevertheless, it has to be pointed out that these myeloid cell types usually present a high degree of plasticity. For example, macrophages and DCs can differentiate from monocytes. These cell types will be described briefly as follows.

2.2.1.1   Monocytes

They represent between 2 % and 10 % of haematopoietic-derived cells. Monocytes circulate systemically and play important roles in replenishing tissue-resident macrophages. Although monocytes have been classically considered "only the precursors" of macrophages, this view is certainly highly simplistic. Monocytes quickly respond to infection, and enhance inflammation, tissue repair, and eliminate pathogens through phagocytosis. These cells have a life span of about 2 days. Currently, murine monocytes are divided in two categories according to their surface phenotype. The Ly6C^{high} monocyte subset (also known as inflammatory monocytes) is mainly involved in inflammation and antimicrobial immune responses, while the Ly6C$^{low/neg}$ subset (also known as "patrolling monocytes") is rather a tissue-infiltrating monocyte involved in tissue repair. In fact, monocytes are quite capable of microbe phagocytosis and killing through production of reactive oxygen species (ROS) following the respiratory burst. They can also trigger (in some circumstances) T cell responses, and polarize immune responses through cytokine secretion. Phenotypically, both subsets express the myeloid lineage marker CD11b, together with CD115, F4/80^{low}, and MHC II [1, 2]. Importantly, the Ly6C^{high} subset lacks expression of CD62L while the Ly6C^{low} subset expresses this homing surface molecule [3]. Monocytes do not express Ly6G or CD11c.

2.2.1.2 Macrophages

Macrophages comprise a very heterogeneous, highly phagocytic cell population. These cells are further classified into several "subclasses" according to ontology, tissue localization, and specialized functions. Thus, we can have osteoclasts, microglia, Kupffer cells, alveolar macrophages, and ubiquitous tissue-resident macrophages. While it was previously thought that macrophages differentiated from bone marrow-derived monocytes, it has been shown that adult macrophages arise from at least two other different sources, one of these being embryonic progenitors [4, 5], and the other from tissue-resident precursors. Phenotypically, macrophages are similar to other myeloid cells, characterized as CD11b$^+$ CD11c$^{low/neg}$, F4/80$^{+/}$ ^{high}, MHC II$^+$, and CD68$^+$ [6, 7]. Most macrophages also express IL4Rα, CD163 and they can also express Ly6C to varying levels. Nevertheless, murine macrophages are Ly6G$^{neg/low}$ and they are most frequently identified by high expression levels of CD11b, F4/80, and CD68 [8].

2.2.1.3 Dendritic Cells (DCs)

Dendritic cells are probably the most immunogenic myeloid cell lineage, and they play a fundamental role in linking innate with adaptive immunity [9–12]. DCs are very efficiently activated through the recognition of pathogen-derived and danger molecules, leading to strong up-regulation of T cell co-stimulatory molecules. Therefore, they are potent activators of naive T cells. DCs are roughly classified in two main groups; conventional (or myeloid) DCs, and plasmacytoid DCs [13]. Here we will focus on conventional DCs. The phenotype of DCs is certainly very plastic, and depends on their anatomical localization and their maturation degree. Nevertheless, they are frequently identified as CD11b$^+$, CD11c^{high}, and MHC II$^+$ cells. They lack Ly6G and F4/80 expression, and they can express Ly6C at varying degrees. DC-SIGN and CD123 are also additional good markers frequently used to differentiate DCs from macrophages.

2.2.1.4 Granulocytes

Granulocytes are short-lived, highly cytotoxic myeloid cells with very high phagocytic activities. They are produced from bone marrow precursors at very large numbers, and are subdivided in three classes: neutrophils, basophils, and eosinophils. This classification is based on the staining properties of their numerous cytoplasmic granules. Possibly, the most common markers for their differentiation from other myeloid lineages are CD11b and GR-1^{high}. The GR-1 epitope is present on two surface molecules, Ly6C and Ly6G. Granulocytes express high levels of Ly6G. Apart from these molecules, granulocytes also express CD62L at varying levels. They are also negative for CD11c, a useful marker to discriminate them from DCs. Myeloperoxidase (MPO) is also used as a granulocyte (neutrophil) marker, although this is still controversial as it might also be expressed in monocytes and macrophages [14].

2.2.2 Physiological Myelopoiesis

Haematopoiesis is a highly regulated process absolutely necessary to keep the homeostasis of the whole body. Thanks to this process, erythrocytes, platelets, and immune cells are continuously produced to meet the demands of the organism. This process entirely relies on a relatively low number of pluripotent haematopoietic stem cells (HSCs, Lin^{neg}, Sca-1$^+$, and cKit$^+$; also called LSK stage) which are kept most of the time quiescent within a specialized niche within the bone marrow (Fig. 2.1). LSK cells are maintained in the presence of stem cell factor (SCF) and leukemia inhibitory factor (LIF), mainly. The differentiation from HSCs toward myeloid cells is usually thought to be a sequentially-regulated pathway through different intermediate differentiation stages (Fig. 2.1) [15, 16]. Stromal cells and other cell types produce cytokines that mobilize these HSCs to differentiate in all the variety of blood cells. The specific site and cytokine combination will lead to the differentiation of each cell lineage. Within the LSK population, a CD41^{high} subset is committed toward myeloid–erythroid differentiation [17]. These cells then give rise to the common myeloid progenitor mainly by the activities of GM-CSF and SCF (CMP, Lin^{neg} Sca-1^{neg} cKit$^+$ CD41$^{low/neg}$ CD64^{low} CD34$^+$ CD115^{neg}). Other cytokines also contribute to their differentiation, including IL3, IL6, and Flt3L. CMPs further differentiate into the granulocyte/monocyte progenitor (GMP, Lin^{neg} Sca-1^{neg} cKit$^+$ CD64^{high} CD34$^+$). GM-CSF, IL3, and macrophage colony-stimulating factor (M-CSF) drive further their differentiation toward monocytes, macrophages, and DCs through a common progenitor termed the "monocyte-macrophage-DC progenitor" or MDP (CD11b^{neg}, Ly6C^{neg}, CD117$^+$, CD115$^+$ CD135$^+$). MDPs give rise

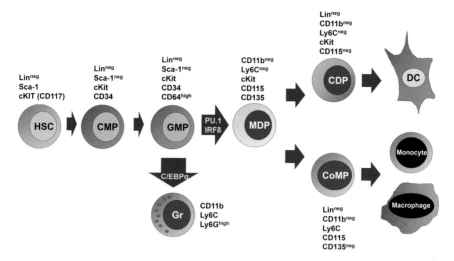

Fig. 2.1 Physiological myelopoiesis. A simplified scheme of myelopoiesis in physiological conditions. Each intermediate precursor is shown with the most characteristic phenotype above. The key transcription factors regulating lineage commitment are shown within the appropriate arrows. *HSC* haematopoietic stem cell; *CMP* common myeloid precursor; *GMP* granulocyte–monocyte precursor; *MDP* monocyte–macrophage-DC precursor; *CoMP* common monocyte–macrophage precursor; *CDP* common DC progenitor; *DC* dendritic cell; *Gr* granulocyte

to at least two distinct committed myeloid precursors, the common DC progenitor (CDP) and the common monocyte progenitor (CoMP, Lin^{neg}, CD11b^{neg}, Ly6C$^+$, CD115$^+$, CD135^{neg}). Importantly, the CDP gives rise to both conventional and plasmacytoid DCs [13, 18], while the CoMP leads to monocyte–macrophage differentiation [2]. Granulocytes are differentiated from the GMP mainly by the activity of granulocyte colony-stimulating factor.

Myelopoiesis can also take place extramedullary, in the spleen where myeloid progenitors are present and can differentiate into several myeloid lineages including monocytes and DCs [2, 19].

2.2.3 Transcription Factors Driving Myelopoiesis

Transcriptionally, myelopoiesis is also regulated by the coordinated expression of a few key transcription factors that shift the commitment of precursors toward monocyte, macrophage, or granulocyte differentiation. HSCs require the expression of PU.1 and GABP to differentiate toward the myeloid lineages [20]. Runx1 leads to C/EBPα expression [21], and then a regulated interplay of C/EBP-AP1 transcription factors determines monocyte or granulocyte differentiation [22]. At high AP-1:C/EBP ratios, these factors heterodimerize leading to monocytes. However, the formation of C/EBPα homodimers favors granulocyte differentiation. IRF8 expression is also critical for monocyte/DC differentiation, as it blocks C/EBP activities and inhibits granulocyte (neutrophil) differentiation [23]. Additionally, its inactivation by p38-mediated phosphorylation is enough to inhibit neutrophil development [24]. Monocyte differentiation also seems to be controlled by NFAT through ERK-dependent activities, while granulocyte differentiation was shown to depend on PU.1 and STAT3 through the participation of JAK1 and calmodulin kinase II [25]. It is important to mention that steady-state haematopoiesis can be significantly altered under certain circumstances, resulting in nonsteady-state hae-matopoiesis. This is the result of circumstances such as infection or cytokine stimulation. In this situation, large numbers of neutrophils are produced through the activities of C/EBPβ [26]. Significantly, C/EBPβ is also highly induced in cancer-bearing patients [27].

2.3 MDSC Differentiation in Cancer

2.3.1 Myeloid Cells and Cancer Progression in Murine Models

Since the early 70s, it was observed that tumor-bearing patients exhibited a sys-temic increase in myeloid cells, particularly cells resembling neutrophils (neutro-phylia) [28]. Tumor infiltration by myeloid cells also correlated with tumor progression, metastasis, and poor prognosis. Although most studies focused on tumor-infiltrating macrophages and DCs, there was a significant population of cells that did not fit within these myeloid populations. These cells were highly sup-pressive and exhibited a markedly immature phenotype, as they did not express high levels of maturation markers and MHC molecules characteristically found in classical myeloid cells [28, 29]. Removal of tumors reverted to the numbers of circulating myeloid cells and neutrophylia, suggesting that tumors were directly producing factors leading to the expansion of these cells.

2.3.2 The Definition of Myeloid-Derived Suppressor Cells

These cancer-related immunosuppressive cells exhibited a variety of phenotypes, and they seemed to correspond to a heterogenous population of cells of myeloid origin. These cells were known by several nomenclatures which included "null cells", "immature suppressive cells" or they were even confounded with tumor-infiltrating macrophages or tolerogenic DCs. Studies in murine cancer models showed that these cells highly co-expressed CD11b and GR1, and could be further classified into two subtypes according to the expression of Ly6C and Ly6G [30]; monocytic $Ly6C^{high}Ly6G^{neg/low}$ M-MDSCs, and granulocytic $Ly6C^{low}Ly6G^{high}$ G-MDSCs. As monocytic and granulocytic MDSCs exhibited mono- or polymor-phonuclei, respectively, many researchers concluded that these cell subsets were either dysfunctional inflammatory monocytes or tumor-associated neutrophils [31, 32]. Nevertheless, although their phenotypes might resemble those of monocytes and neutrophils, their functional differences suggested that these cells could either represent alternative functional states of these myeloid cells, or arise independently from monocytes and neutrophils through a perturbed myelopoiesis [30, 33]. Thus, M-MDSCs and G-MDSCs were considered at first dysfunctional immature myeloid cells that arose independently from each other.

Consequently, to avoid misunderstandings, the term "myeloid-derived suppressor cell" was coined by a group of researchers that pioneered research in MDSCs [34]. Although there is still some controversy on terminology, the MDSC term is proving useful from a practical point of view until the MDSC ontology is unambiguously unmasked.

2.3.3   Perturbed Myelopoiesis Behind MDSC Differentiation

Whether MDSCs are considered a "dysfunctional state" of monocytes/neutrophils, or myeloid lineages that arise independently, it is evident that perturbed myelopoiesis caused by cancer is behind MDSC differentiation and expansion. Growing tumors secrete a wide range of cytokines and metabolites that distribute systemically through circulation, also reaching the bone marrow. Some of these secreted factors have been identified in vitro in cultures of a wide range of cancer cell lines. Not surprisingly, they comprise of a collection of molecules that drive myeloid differentiation and include cytokines such as GM-CSF, G-CSF, M-CSF, IL6, IL13, IL4, and SCF [27, 30, 35, 36]. Other molecules have also been shown to contribute to MDSC differentiation and acquisition of immunosuppressive functions, such as prostaglandin E_2, TGF-β, and vasoactive intestinal peptide [37–40]. Recently, it was shown that IL18 increases the differentiation of M-MDSCs from CD11b^{neg} precursors [41]. Thus, the significant increase in levels of these circulating molecules and tumor-derived exosomes perturbs myelopoiesis leading to the mobilization of MDSCs [42] (Fig. 2.2).

2.3.4   Do MDSCs Belong to a Specific Myeloid Lineage? Relationship Between M-MDSCs and G-MDSCs

So far, there has been a wide assumption within the scientific community that M-MDSCs and G-MDSCs are largely unrelated. Possibly, their resemblance to inflammatory monocytes and neutrophils, respectively, provides weight to this assumption. However, a direct ontological relationship between monocytes with M-MDSCs or neutrophils with G-MDSCs has not been shown yet. In fact, morphologically and phenotypically, M-MDSCs do resemble monocytes. M-MDSCs are mononuclear cells which express CD11b, Ly6C^{high}, Ly6G^{low}, CD62L, CD115,

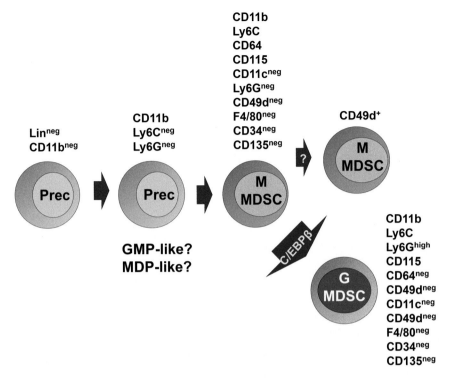

Fig. 2.2 Putative MDSC differentiation pathway. A simplified scheme of MDSC differentiation from putative precursors (Prec) is shown in the figure. So far, the specific nature and phenotype of these precursors are unknown, but they are likely to differ from those involved in physiological myelopoiesis. In this scheme, monocytic MDSCs (M-MDSCs) are precursors of granulocytic MDSCs (G-MDSC). The known phenotype of each MDSC subset is indicated in the figure. The participation in MDSC differentiation of the transcription factor C/EBPβ is indicated, although it is likely that this transcription factor is acting at the level of precursors

and CD64a/b [43, 44]. The surface marker CD115 (M-CSF receptor) is expressed early in haematopoiesis at least from the MDP progenitor stage, and this includes all the monocytic progeny. CD64a/b is constitutively expressed by monocytes, macrophages, and DCs [45]. CD49d which is expressed in a highly suppressive M-MDSC subset is also expressed by monocytes, macrophages, and DCs. Therefore, M-MDSCs are apparently "monocytes", or at least of monocytic origin. In physiological haematopoiesis, granulocytes arise from a GMP stage, expressing their "hallmark" surface marker GR1. The GR1 epitope is present on two surface molecules, Ly6C and Ly6G. Similarly to inflammatory neutrophils, G-MDSCs are Ly6G^{high} cells. Several studies claim that circulating M-MDSCs can infiltrate tumors, and then under the influence of tumor-derived factors, they differentiate toward tumor-associated macrophages, DCs, and neutrophils-G-MDSCs [46, 47]. Some authors consider G-MDSCs as tumor-infiltrating neutrophils [48]. In fact, transcriptomic analyses between neutrophils and G-MDSCs showed that they are similar cells, although there were significant differences in the expression of key functional enzymes such as lysosomal proteins, arginase, myeloperoxidase, and production of reactive oxygen species [49]. A second transcriptomic study between naïve neutrophils, splenic G-MDSCs, and tumor-infiltrating neutrophils showed that the mRNA profiles of the three cell populations were significantly different, but MDSCs were more similar to naïve neutrophils [50]. Even so, considering all the available experimental evidence it seems that G-MDSCs are not truly *bona fide* neutrophils [33]. In agreement with the inhibitory role of IRF8 on neutrophil development, IRF8 expression also suppresses MDSC differentiation [51]. However, there is a fundamental difference; as described above, for physiological neutrophil differentiation the expression of C/EBP-α is required [24]. In contrast, the C/EBP-β isoform, which is associated to nonsteady-state haematopoiesis [26] is responsible for driving MDSC differentiation [27].

According to physiological haematopoiesis, monocytes and granulocytes share a common progenitor early in the granulocyte–monocyte progenitor stage. Following this "commitment model", it seems unlikely that inflammatory monocytes in steady-state conditions may differentiate into neutrophils. Their differentiation is antagonistic and controlled by IRF8 expression. However, there is some evidence of transdifferentiation from inflammatory neutrophils into monocyte/macrophages under inflammatory conditions through the activity of p38 [52].

Surprisingly, purified M-MDSCs quickly differentiate towards G-MDSCs in vitro, ascertained by strong Ly6G up-regulation [44]. Moreover, the same phenomenon was observed in vivo, where M-MDSCs differentiated towards G-MDSCs after infiltrating tumors [53]. Therefore, these results strongly suggest that M-MDSCs and G-MDSCs are directly related rather than being independent myeloid lineages arising from unrelated pathways within the bone marrow. Thus, as the MDSC precursor cell possesses monocytic markers but acquires granulocytic markers toward terminal differentiation, it is highly likely that MDSC subsets directly derive from GMPs.

2.4 Summary and Conclusions

Growing tumors strongly alter physiological myelopoiesis leading to the differentiation and expansion of MDSCs. These cells clearly belong to the myeloid lineage, although their discrimination from "physiological" myeloid cell types is rather challenging. Nevertheless, cancer is a rather unusual pathology, and it is highly likely that the same standards for classifying myeloid cells in physiological conditions do not apply in pathological situations. Classically, MDSCs have been classified into two subsets according to their phenotype; monocytic and granulocytic MDSCs, phenotypically resembling inflammatory monocytes and neutrophils, respectively. As differentiation of granulocytic and monocytic lineages in physiological conditions seems to be antagonistic and dependent on the expression levels of IRF8, it has been assumed that M-MDSC and G-MDSC differentiation pathways are rather independent from each other. However, there is compelling evidence that MDSCs derive from a $CD11b^+$ $Ly6C^{neg}$ $Ly6G^{neg}$ precursor leading to M-MDSCs. Then, these M-MDSCs within the tumor environment further differentiate to G-MDSCs which possess a relatively short life. Therefore, rather than a "heterogeneous population of immature myeloid precursors", MDSCs should be considered as an alternative "myeloid lineage" that appears in pathological conditions. This MDSC "cell type" is expanded at large levels and it is systemically present at different differentiation degrees, being the G-MDSC the terminal differentiation stage.

Acknowledgments David Escors is funded by a Miguel Servet Fellowship (CP12/03114), a FIS project grant (PI14/00579) from the Instituto de Salud Carlos III, Spain, the Refbio transpyrenaic collaborative project grants (NTBM), a Sandra Ibarra Foundation grant, Gobierno de Navarra Grant (BMED 033-2014), and a Gobierno Vasco BioEf project grant (BIO13/CI/014).

References

1. Shi C, Pamer EG (2011) Monocyte recruitment during infection and inflammation. Nat Rev Immunol 11(11):762–774. doi:10.1038/nri3070
2. Hettinger J, Richards DM, Hansson J, Barra MM, Joschko AC, Krijgsveld J, Feuerer M (2013) Origin of monocytes and macrophages in a committed progenitor. Nat Immunol 14(8):821–830. doi:10.1038/ni.2638
3. Italiani P, Boraschi D (2014) From monocytes to M1/M2 macrophages: phenotypical versus Functional differentiation. Front Immunol 5:514. doi:10.3389/fimmu.2014.00514
4. Epelman S, Lavine KJ, Beaudin AE, Sojka DK, Carrero JA, Calderon B, Brija T, Gautier EL, Ivanov S, Satpathy AT, Schilling JD, Schwendener R, Sergin I, Razani B, Forsberg EC, Yokoyama WM, Unanue ER, Colonna M, Randolph GJ, Mann DL (2014) Embryonic and adult-derived resident cardiac macrophages are maintained through distinct mechanisms at steady state and during inflammation. Immunity 40(1):91–104. doi:10.1016/j.immuni.2013.11.019
5. Epelman S, Lavine KJ, Randolph GJ (2014) Origin and functions of tissue macrophages. Immunity 41(1):21–35. doi:10.1016/j.immuni.2014.06.013
6. Yang J, Zhang L, Yu C, Yang XF, Wang H (2014) Monocyte and macrophage differentiation: circulation inflammatory monocyte as biomarker for inflammatory diseases. Biomark Res 2 (1):1. doi:10.1186/2050-7771-2-1
7. Murray PJ, Wynn TA (2011) Protective and pathogenic functions of macrophage subsets. Nat Rev Immunol 11(11):723–737. doi:10.1038/nri3073
8. Zhang X, Goncalves R, Mosser DM (2008) The isolation and characterization of murine macrophages. Curr Protoc Immunol Chapter 14: Unit 14 11. doi:10.1002/0471142735.im1401s83
9. Breckpot K, Escors D (2009) Dendritic cells for active anti-cancer immunotherapy: targeting activation pathways through genetic modification. Endocr Metab Immune Disord Drug Targets 9:328–343
10. Goold HD, Escors D, Conlan TJ, Chakraverty R, Bennett CL (2011) Conventional dendritic cells are required for the activation of helper-dependent CD8 T cell responses to a model antigen after cutaneous vaccination with lentiviral vectors. J Immunol 186(8):4565–4572
11. Steinman RM, Cohn ZA (1973) Identification of a novel cell type in peripheral lymphoid organs of mice. I. Morphology, quantitation, tissue distribution. J Exp Med 137(5):1142–1162

12. Steinman RM, Banchereau J (2007) Taking dendritic cells into medicine. Nature 449 (7161):419–426. doi:10.1038/nature06175
13. Onai N, Kurabayashi K, Hosoi-Amaike M, Toyama-Sorimachi N, Matsushima K, Inaba K, Ohteki T (2013) A clonogenic progenitor with prominent plasmacytoid dendritic cell developmental potential. Immunity 38(5):943–957. doi:10.1016/j.immuni.2013.04.006
14. Amanzada A, Malik IA, Nischwitz M, Sultan S, Naz N, Ramadori G (2011) Myeloperoxidase and elastase are only expressed by neutrophils in normal and in inflamed liver. Histochem Cell Biol 135(3):305–315. doi:10.1007/s00418-011-0787-1
15. Akashi K, Traver D, Miyamoto T, Weissman IL (2000) A clonogenic common myeloid progenitor that gives rise to all myeloid lineages. Nature 404(6774):193–197. doi:10.1038/35004599
16. Nishikawa S, Goldstein RA, Nierras CR (2008) The promise of human induced pluripotent stem cells for research and therapy. Nat Rev Mol Cell Biol 9(9):725–729. doi:10.1038/nrm2466
17. Miyawaki K, Arinobu Y, Iwasaki H, Kohno K, Tsuzuki H, Lino T, Shima T, Kikushige Y, Takenaka K, Miyamoto T, Akashi K (2015) CD41 marks the initial myelo-erythroid lineage specification in adult mouse hematopoiesis: redefinition of murine common myeloid progenitor. Stem Cells (Dayton, Ohio) 33(3):976–987
18. Onai N, Obata-Onai A, Schmid MA, Ohteki T, Jarrossay D, Manz MG (2007) Identification of clonogenic common Flt3+M-CSFR+ plasmacytoid and conventional dendritic cell progenitors in mouse bone marrow. Nat Immunol 8(11):1207–1216. doi:10.1038/ni1518
19. Arce F, Rowe HM, Chain B, Lopes L, Collins MK (2009) Lentiviral vectors transduce proliferating dendritic cell precursors leading to persistent antigen presentation and immunization. Mol Ther 17(9):1643–1650
20. Yang ZF, Drumea K, Cormier J, Wang J, Zhu X, Rosmarin AG (2011) GABP transcription factor is required for myeloid differentiation, in part, throught its control of Gfi-1 expression. Blood 118(8):2243–2253
21. Friedman AD (2015) C/EBPalpha in normal and malignant myelopoiesis. Trans Int J Hematol 101(4):330–341
22. Hong S, Skaist AM, Wheelan SJ, Friedman AD (2011) AP-1 protein induction during monopoiesis favors C/EBP:AP-1 heterodimers over C/EBP homodimerization and stimulates FosB transcription. J Leukoc Biol 90(4):643–651
23. Tamura T, Kurotaki D, Koizumi S (2015) Regulation of myelopoiesis by the transcription factor IRF8. Int J Hematol 101(4):342–351
24. Geest CR, Buitenhuis M, Laarhoven AG, Bierings MB, Bruin MC, Vellenga E, Coffer PJ (2009) p38 MAP kinase inhibits neutrophil development through phosphorylation of C/EBPalpha on serine 21. Stem Cells (Dayton, Ohio) 27(9):2271–2282
25. Barbosa CM, Bincoletto C, Barros CC, Ferreira AT, Paredes-Gamero EJ (2014) PLCgamma2 and PKC are important to myeloid lineage commitment triggered by M-SCF and G-CSF. J Cell Biochem 115(1):42–51
26. Hirai H, Yokota A, Tamura A, Sato A, Maekawa T (2015) Non-steady-state hematopoiesis is regulated by the C/EBPbeta transcription factor. Cancer Sci doi:10.1111/cas.12690
27. Marigo I, Bosio E, Solito S, Mesa C, Fernandez A, Dolcetti L, Ugel S, Sonda N, Bicciato S, Falisi E, Calabrese F, Basso G, Zanovello P, Cozzi E, Mandruzzato S, Bronte V (2010) Tumor-induced tolerance and immune suppression depend on the C/EBPbeta transcription factor. Immunity 32(6):790–802. doi:10.1016/j.immuni.2010.05.010 S1074-7613(10)00202-5 [pii]
28. Talmadge JE, Gabrilovich DI (2013) History of myeloid-derived suppressor cells. Nat Rev 13 (10):739–752
29. Gabrilovich DI, Nagaraj S (2009) Myeloid-derived suppressor cells as regulators of the immune system. Nat Rev Immunol 9(3):162–174. doi:10.1038/nri2506 nri2506 [pii]
30. Youn JI, Nagaraj S, Collazo M, Gabrilovich DI (2008) Subsets of myeloid-derived suppressor cells in tumor-bearing mice. J Immunol 181(8):5791–5802

31. Lutz MB, Kukutsch NA, Menges M, Rossner S, Schuler G (2000) Culture of bone marrow cells in GM-CSF plus high doses of lipopolysaccharide generates exclusively immature dendritic cells which induce alloantigen-specific CD4 T cell anergy in vitro. Eur J Immunol 30 (4):1048–1052

32. Pillay J, Kamp VM, van Hoffen E, Visser T, Tak T, Lammers JW, Ulfman LH, Leenen LP, Pickkers P, Koenderman L (2012) A subset of neutrophils in human systemic inflammation inhibits T cell responses through Mac-1. J Clin Invest 122(1):327–336. doi:10.1172/JCI57990 57990 [pii]

33. Tsiganov EN, Verbina EM, Radaeva TV, Sosunov VV, Kosmiadi GA, Nikitina IY, Lyadova IV (2014) Gr-1dimCD11b$^+$ immature myeloid-derived suppressor cells but not neutrophils are markers of lethal tuberculosis infection in mice. J Immunol 192(10): 4718–4727. doi:10.4049/jimmunol.1301365

34. Gabrilovich DI, Bronte V, Chen SH, Colombo MP, Ochoa A, Ostrand-Rosenberg S, Schreiber H (2007) The terminology issue for myeloid-derived suppressor cells. Cancer Res 67(1):425; author reply 426. doi:67/1/425 [pii] 10.1158/0008-5472.CAN-06-3037

35. Morales JK, Kmieciak M, Knutson KL, Bear HD, Manjili MH (2010) GM-CSF is one of the main breast tumor-derived soluble factors involved in the differentiation of CD11b-Gr1- bone marrow progenitor cells into myeloid-derived suppressor cells. Breast Cancer Res Treat 123 (1):39–49

36. Highfill SL, Rodriguez PC, Zhou Q, Goetz CA, Koehn BH, Veenstra R, Taylor PA, Panoskaltsis-Mortari A, Serody JS, Munn DH, Tolar J, Ochoa AC, Blazar BR (2010) Bone marrow myeloid-derived suppressor cells (MDSCs) inhibit graft-versus-host disease (GVHD) via an arginase-1-dependent mechanism that is up-regulated by interleukin-13. Blood 116 (25):5738–5747. doi:10.1182/blood-2010-06-287839 blood-2010-06-287839 [pii]

37. Valenti R, Huber V, Filipazzi P, Pilla L, Sovena G, Villa A, Corbelli A, Fais S, Parmiani G, Rivoltini L (2006) Human tumor-released microvesicles promote the differentiation of myeloid cells with transforming growth factor-beta-mediated suppressive activity on T lymphocytes. Cancer Res 66(18):9290–9298

38. Valenti R, Huber V, Iero M, Filipazzi P, Parmiani G, Rivoltini L (2007) Tumor-released microvesicles as vehicles of immunosuppression. Cancer Res 67(7):2912–2915

39. Obermajer N, Muthuswamy R, Lesnock J, Edwards RP, Kalinski P (2013) Positive feedback between PGE2 and COX2 redirects the differentiation of human dendritic cells toward stable myeloid-derived suppressor cells. Blood 118(20):5498–5505

40. Li G, Wu K, Tao K, Lu X, Ma J, Mao Z, Li H, Shi L, Li J, Niu Y, Xiang F, Wang G (2015) Vasoactive intestinal peptide induces CD14 + HLA-DR-/low myeloid-derived suppressor cells in gastric cancer. Mol Med Rep 12(1):760–768

41. Lim HX, Hong HJ, Cho D, Kim TS (2014) IL-18 enhances immunosuppressive responses by promoting differentiation into monocytic myeloid-derived suppressor cells. J Immunol 193 (11):5453–5460

42. Dolcetti L, Peranzoni E, Ugel S, Marigo I, Fernandez Gomez A, Mesa C, Geilich M, Winkels G, Traggiai E, Casati A, Grassi F, Bronte V (2010) Hierarchy of immunosuppressive strength among myeloid-derived suppressor cell subsets is determined by GM-CSF. Eur J Immunol 40(1):22–35. doi:10.1002/eji.200939903

43. Huang B, Pan PY, Li Q, Sato AI, Levy DE, Bromberg J, Divino CM, Chen SH (2006) Gr-1+ CD115+ immature myeloid suppressor cells mediate the development of tumor-induced T regulatory cells and T-cell anergy in tumor-bearing host. Cancer Res 66(2):1123–1131

44. Liechtenstein T, Perez-Janices N, Gato M, Caliendo F, Kochan G, Blanco-Luquin I, Van der Jeught K, Arce F, Guerrero-Setas D, Fernandez-Irigoyen J, Santamaria E, Breckpot K, Escors D (2014) A highly efficient tumor-infiltrating MDSC differentiation system for discovery of anti-neoplastic targets, which circumvents the need for tumor establishment in mice. Oncotarget 5(17):7843–7857

45. Liechtenstein TM (2015) Lentivector-based cancer immunotherapy silencing PD-L1 and modulating cytokine priming; development of ex vivo myeloid-derived suppressor cells to assess therapeutic efficacy. PhD thesis. University College London, London

46. Kusmartsev S, Gabrilovich DI (2005) STAT1 signaling regulates tumor-associated macrophage-mediated T cell deletion. J Immunol 174(8):4880–4891
47. Corzo CA, Condamine T, Lu L, Cotter MJ, Youn JI, Cheng P, Cho HI, Celis E, Quiceno DG, Padhya T, McCaffrey TV, McCaffrey JC, Gabrilovich DI (2010) HIF-1alpha regulates function and differentiation of myeloid-derived suppressor cells in the tumor microenvironment. J Exp Med 207(11):2439–2453. doi:10.1084/jem.20100587 jem. 20100587 [pii]
48. Rodriguez PC, Ernstoff MS, Hernandez C, Atkins M, Zabaleta J, Sierra R, Ochoa AC (2009) Arginase I-producing myeloid-derived suppressor cells in renal cell carcinoma are a subpopulation of activated granulocytes. Cancer Res 69(4):1553–1560. doi:10.1158/0008-5472.CAN-08-1921
49. Youn JI, Collazo M, Shalova IN, Biswas SK, Gabrilovich DI (2012) Characterization of the nature of granulocytic myeloid-derived suppressor cells in tumor-bearing mice. J Leukoc Biol 91(1):167–181. doi:10.1189/jlb.0311177
50. Fridlender ZG, Sun J, Mishalian I, Singhal S, Cheng G, Kapoor V, Hornq W, Fridlender G, Bayuh R, Worthen GS, Albelda SM (2012) Transcriptomic analysis comparing tumor-associated neutrophils with granulocytic myeloid-derived suppressor cells and normal neutrophils. PLoS ONE 7(2):e31524
51. Waight JD, Netherby C, Hensen ML, Miller A, Hu Q, Liu S, Bogner PN, Farren MR, Lee KP, Liu K, Abrams SI (2013) Myeloid-derived suppressor cell development is regulated by a STAT/IRF-8 axis. J Clin Invest 123(10):4464–4478
52. Koffel R, Meshcheryakova A, Warszawska J, Henning A, Wagner K, Jorgl A, Gubi D, Moser D, Hladik A, Hoffmann U, Fischer MB, van der Berg W, Koenders M, Scheinecker C, Gesslbauer B, Knapp S, Strobl H (2014) Monocytic cell differentiation from band-stage neutrophils under inflammatory conditions via MKK6 activation. Blood 124(17):2713–2724
53. Youn JI, Kumar V, Collazo M, Nefedova Y, Condamine T, Cheng P, Villagra A, Antonia S, McCaffrey JC, Fishman M, Sarnaik A, Horna P, Sotomayor E, Gabrilovich DI (2013) Epigenetic silencing of retinoblastoma gene regulates pathologic differentiation of myeloid cells in cancer. Nat Immunol 14(3):211–220. doi:10.1038/ni.2526 ni 2526 [pii]

Chapter 3
Human MDSCs

Grazyna Kochan

Abstract Myeloid-derived suppressor cells strongly expand in many pathological conditions including cancer, and they suppress immunological responses by interfering with the effector functions of T cells, dendritic cells, and NK cells. The differentiation and accumulation of MDSCs is a negative outcome caused by the interplay between tumor cells and myelopoiesis. Since the phenotype of MDSCs and their mechanisms of action seem to depend on the type of cancer and stage of the disease, it is important to evaluate which MDSC subsets have prognostic values in the outcome of the disease. In the present chapter we will systematize the current information on the different populations of human MDSCs and their markers as well as their similarities/differences with MDSCs from murine models.

Keywords Human MDSC · Myeloid differentiation · Cancer · MDSC phenotype

3.1 Introduction

Although the participation of myeloid cells in cancer progression was known since 1960s, MDSCs have been extensively studied in these recent years. MDSC populations are of high interest in oncology, as the expansion of these cells in cancer is significantly elevated. There is an increasing amount of evidence that their suppressive activity correlates with negative prognosis and poor overall survival (OS) in cancer patients. Accumulation of MDSCs is attributed to tumor progression and thus, the presence of MDSCs was proposed as a potential biomarker associated to disease progression and OS of patients.

However, the major problem in using MDSCs as a prognostic biomarker and their utilization in cancer research is in fact the difficulty in defining the MDSCs themselves, especially in humans. This is a key factor to differentiate pathological

G. Kochan (✉)
Navarrabiomed-Biomedical Research Centre, Fundación Miguel Servet, IdiSNA,
Calle Irunlarrea 3, Pamplona, Navarra 31008, Spain
e-mail: grkochan@navarra.es

MDSCs from other physiological myeloid cell types. These cells have been characterized in different tumors and defined by various phenotypes. Nevertheless, the study of MDSC surface markers leads to the conclusion that their phenotype strongly depends on the tumor type and developmental stage of the disease.

In the case of human cancers, it is quite difficult to define MDSCs, as a growing number of different phenotypes determined in different tumor types appear in the specialized literature. An additional difficulty comes from the fact that the biological material for analyses comes from different stages of the disease, distinct cancer backgrounds, and from patients receiving anticancer treatments and most of the times from peripheral blood instead of the tumor itself.

The characteristics of murine MDSCs (described here in detail in other chapters) are much simpler and allows to divide them into two well-defined phenotypical types: monocytic and granulocytic MDSCs. Moreover, it has been shown by two independent research groups that granulocytic MDSCs comprise of the final maturation stage of monocytic MDSCs [1, 2]. A brief description of murine MDSCs and theirs markers is provided below.

3.2 Murine MDSCs

Early works during the late 1990s in murine tumor models defined MDSCs by a specific phenotype based on the expression of CD11b and Gr-1 markers [3, 4]. Some of the initial observations on the accumulation of these cells that inhibited $CD8^+$ cytotoxic lymphocytes in immunocompetent hosts were made by Bronte and colleagues during the work on therapeutic anticancer vaccines [3, 5]. These cells were further studied and additional markers such as Ly6G, Ly6C, and interleukin $4R\alpha$ were identified. However, MDSCs were soon found to be heterogeneous population, although they could still be further classified in two main subsets based on the basis of expression of other surface markers: monocytic Mo-MDSCs characterized with the phenotype $CD11b^+$ $Gr-1^{int}$ $Ly-6C^{hi}$ $Ly-6G^-$ and granulocytic G-MDSC (PMN-MDSC) $CD11b^+$ $Gr-1^{hi}$ $Ly-6C^{low}$ $Ly-6G^+$ [6]. Greten and colleagues showed that monocytic MDSCs could also be divided into $CD49d^+$ and $CD49d^-$ subsets, of which $CD49^-$ Mo-MDSCs were the strongest T-cell suppressors, even more than G-MDSCs.

Those two subsets are found in all murine tumor models, but the proportion of Mo-MDSC and G-MDSC varies in different tumor models. There are conflicting results on which subset is dominant in peripheral lymphoid organs and within the tumor itself. Relative percentage differences between these two subsets may vary significantly from one cancer type to another. This can be explained by the fact that there is evidence that G-MDSC is the mature stage of Mo-MDSC. Therefore, the different tumor environments (pH hypoxia, metabolic products, and different tumor origins) possibly influence the maturation kinetics in different ways.

3.3 Human MDSCs

Compared to the murine system, available studies on the characterization of human tumor-infiltrating MDSC are still very limited. One explanation is the restriction on the availability of biological samples. In addition, the human MDSC phenotype seems to be different from their murine counterparts. As human MDSCs lack a homologue molecule to the murine Gr-1, a varying combination of markers is being used to define monocytic and granulocytic human MDSC populations. Unfortunately, some of the markers chosen to characterize human MDSCs in fact overlap partially or completely with what are considered as the equivalents to murine Mo-MDSC and G-MDSC populations.

The majority of research groups define the human MDSCs phenotype by the expression of well-characterized myeloid antigens such as CD11b, or CD33 in combination with low or absent HLA-DR expression. Then, an attempt to further classify these cells into monocytic and granulocytic subsets has been carried out on the basis of CD14 and CD15 expression. CD14$^+$ cells are considered monocytic while CD15$^+$ cells would correspond to granulocytic MDSCs. It has also been suggested that additional markers in combination with the ones mentioned above could be more specific and useful to define human MDSCs such as IL4Rα, VEGFR, and CD66b [7, 8]. Both monocytic and granulocytic MDSCs often lack other lineage-specific antigens. In addition, some of the markers mentioned above are not always present in all human MDSCs subsets according to a number of studies, so the best way to definitely define MDSCs is still by the expression of functional markers as ARG1, iNOS, and ROS production. In a significant number of reports and publications, only the phenotypic characterization of MDSCs is reported, without further functional analyses to complement the studies which limit the impact of these reports. This is also explained by the intrinsic difficulty of the "human system", thus a reflection of the limited amount of biological material that is obtained especially from patients [9, 10]. A major drawback on MDSC characterization is that physiological myeloid populations also share all or part of the markers currently used in human MDSC research [7, 11]. Therefore, without functional suppression assays, it is certainly a challenge to identify *bona fide* human MDSCs.

3.4 Other Proposed Human MDSC Phenotypes

Based on the numerous data available in the specialized literature, several other markers have been identified on circulating MDSCs isolated from peripheral blood. However, it is still unclear whether this observed phenotypic diversity is the result of differences in protocols/models of induction and expansion, or just the result of different antibody sets utilized to detect surface markers. In addition to the expression of surface markers, some authors make an emphasis on cell size and granularity as ascertained by flow cytometry, and they have used these criteria to support the phenotype characterization of human MDSCs [12].

While there is a plethora of publications on circulating putative MDSCs from peripheral blood of cancer patients, only a very few studies have been carried out based on the identification of tumor-infiltrating MDSCs. This is in contrast to murine models, in which intra-tumor MDSCs have been studied to a much larger extent. The table below lists some representative examples of the different human MDSC phenotypes identified in studies from a variety of cancer types in human patients. Apart from the phenotypes mentioned here, the literature reports several intermediate phenotypes as the result of the different selection of markers used for analyses.

MDSCs	MDSC phenotype	Tumor	Ref.	
Monocytic	CD14$^+$, HLA-DR$^{low/-}$	Melanoma	[13, 14]	PB,
		SCCHN	[11]	PB, T
	CD14$^+$, IL4Rα, CD124	Melanoma	[7]	PB
		Colon		
	CD33$^+$CD14$^+$, IL4Rα, HLA-DR$^{low/-}$	Glioblastoma	[15]	PB, T
	CD14$^+$, HLA-DR$^{-/low}$, CD11b$^+$, CD33$^+$	Rectal cancer	[16]	PB
Granulocytic	CD15$^+$, IL4Rα$^+$, CD124	Melanoma	[7]	PB, T
		Colon		PB
	Lin$^-$, CD33$^+$, HLA-DR$^{low/-}$CD14$^-$, CD15$^{+,}$ CD11b$^+$	Gastrointestinal	[17]	PB
		Rectal cancer	[16]	PB
	CD33$^+$, CD15$^+$, CD66b$^+$	RCC	[8]	PB
	CD15 + , FSC^{low}, SSC^{high}	RCC		PB
Immature	CD11b$^+$, CD14$^-$, CD15$^-$	Melanoma	[18]	PB, T
	Lin$^{low/-}$, CD33$^+$, CD11b$^+$, HLA-DR$^{low/-}$	Breast cancer	[19]	PB

PB peripheral blood
T tumor

3.5 Mechanisms of Action

Multiple studies have shown that different populations of MDSCs possess distinct suppressive mechanisms and capacities (Fig. 3.1). For example, moMDSC produce arginase (ARG1) that metabolizes arginine to urea and L-ornitine, and iNOS that oxidases L-arginine into citrulline and produces NO. The presence of NO inhibits E-selectin expression on endothelial cells and impairs T-cell recruitment [20]. Depletion of L-arginine also leads to inhibition of CD3ζ expression leading to decreased T-cell proliferation by enforcing a cell cycle arrest in the G_0–G_1 phase [21]. Srivastava and colleagues observed that MDSCs inhibit T cells by cysteine

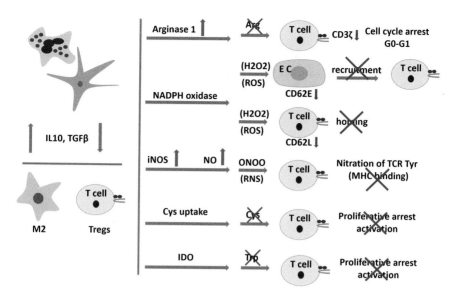

Fig. 3.1 Main immunosuppressive mechanisms of MDSCs. A simplified scheme of the mechanisms used by MDSCs to suppress immune responses. The key factors affecting the immune response are presented in the figure: On the top left, the two subsets of MDSCs are shown (polymorphonuclear granulocytic MDSCs and mononuclear monocytic MDSCs). As shown on the left, MDSC populations exhibit significant cross talk through the expression of IL10 (interleukin 10) and TGFβ (transforming growth factor β) with macrophages (M2, tumor associated macrophages polarized to M2) and regulatory T cells (Tregs); On the right, the main effects over effector T cells and EC (endothelial cells) are indicated; *CD3ζ*, CD3 zeta chain of the T-cell receptor complex; *iNOS* inducible nitric oxide synthetase; *IDO* Indoleamine 2,3-dioxygenase; *CD62* selectin; *MHC* major histocompatibility complex; *Arg* arginine; *Trp* tryptophan; *Cys* cysteine; *NO* nitric oxide; *ROS* reactive oxygen species; *RNS* reactive nitrogen species

uptake that is essential for T-cell activation [22]. Production of ROS leads to generation of peroxynitrite and causes nitration of T-cell receptors resulting in CD8 T-cell suppression [23]. The upregulation of the NADPH oxidase expression in MDSCs also strongly contributes to ROS production.

The main prostaglandin receptor found in MDSC is for PGE$_2$. PGE$_2$ leads to upregulation of arginase and as a result, immune suppression by inhibiting T-cell activities. Obermayer and colleagues observed that PGE$_2$ promotes MDSC recruitment to the tumor environment through induction of stromal cell-derived CXCL12. Granulocytic MDSCs express arginase and myeloperoxidase, producing large quantities of reactive oxygen species (ROS). A study carried out in breast cancer has also shown that IDO upregulation correlates with the immunosuppressive activity of MDSCs [24]. Similar results were obtained from moMDSC in melanoma patients [25]. Novitskiy and colleagues suggested the involvement of IL17 in a mammary carcinoma model as they observed increases in MDSC immunosuppressive function by upregulation arg-1, matrix metalloproteinase-9, IDO, and cyclooxygenase-2 (cox-2), all of them as a result of IL17 action [26].

There is increasing evidence that MDSCs are not only suppressing immune responses in the tumor site but they also significantly contribute to tumor metastasis. It was observed by Yang and colleagues that accumulation of Granulocytic-MDSC in the tumor site was associated to increased bone metastasis [27]. Those MDSC upregulated expression of several MMPs that were critical for the increased invasive capacities of tumor 4T1 cells, both in vitro and in vivo. Different chemokines and chemokine receptors expressed by tumor cells promote MDSC accumulation and as a result of this, there is a rapid tumor growth and dissemination of tumor cells. The role of MDSC in tumor metastasis has been characterized in detail in a review by Condamine et al. [28].

3.6   Models of Human MDSC Differentiation

As mentioned above, one of the complex problems is the identification and characterization of the different subsets of human MDSC. To try to solve this issue, there is an increasing collection of published methods to generate human MDSCs in vitro. It was shown that MDSC (CD11b, or CD33 and HLA-DR$^{low/-}$ CD66b) can be differentiated in vitro from PBMCs by incubation with recombinant GM-CSF and IL6 or GMCSF, IL1β, TNFα, VEGF, and PGE$_2$ [29]. Valenti and colleagues generated CD14$^+$ HLA-DR$^{low/-}$ from monocytes of healthy donors by incubation with IL4 and GM-CSF in the presence of tumor-derived microvesicles [30]. Nevertheless, most published protocols exhibit poor efficiencies of MDSC differentiation. Unlike the murine differentiation systems, most in vitro protocols to derive human MDSCs start from rather differentiated monocytes instead of bone marrow precursors.

These pioneering works represent significant steps forward. However, it is important to establish and standardize protocols to increase the efficiency of in vitro human MDSC differentiation. Additionally, it is yet unclear whether in vitro-generated MDSCs following current protocols are equivalent to tumor-infiltrating subsets. Finally, an efficient system to generate human MDSCs ex vivo will not only help to generate valuable data sets which will ease the characterization of MDSC populations, but would also allow the screening of multiple therapeutic drugs and experimental treatments against them.

3.7   Summary and Conclusions

The difficulty in working with human MDSCs lays in the limited accessibility of material from human tumors. Therefore, the majority of the studies are performed on circulating peripheral blood MDSCs. The identification of numerous populations of MDSCs strongly stresses the necessity of standardization in this field. To help solving this problem, the Association of Cancer Immunotherapy has founded a human MDSC proficiency panel that helps in standardization of immunophenotyping across different MDSC groups.

A proper characterization of MDSC types should lead to a defined panel of antibodies permitting the classification of MDSCs present in the tumor microenvironment. Accurate characterization of these MDSCs would strengthen also the existing few studies on correlation between increased MDSC numbers and subtypes and overall survival index.

Additional in-depth analyses using molecular high throughput technologies (such as genomic, transcriptomic, and proteomic techniques) in the proper MDSC populations would allow the identification of additional key factors and mechanisms responsible for their suppressive activities. High throughput technologies have been already used to generate interactome maps of murine neoplastic MDSCs [31–33].

A way to definitely finish with the current heterogeneity in the classification of human MDSC types could be the use of intra-tumor MDSCs for in-depth phenotypic and mechanistic studies. It has to be taken into account that circulating MDSCs from peripheral blood may include a plethora of myeloid cells at different maturation stages recently released from the bone marrow. Moreover, there are always tumor-associated macrophages, neutrophils, and even basophils that can complicate these analyses. On the other hand, tumor-infiltrating MDSCs probably represent the end stage of MDSC differentiation, being the direct effectors of immunosuppression within the tumor mass. All these considerations suggest that for better understanding of MDSC mechanisms of action more work should be performed on tumor-infiltrating MDSC. While it is challenging to isolate sufficient MDSC numbers from human biopsies, the development of efficient ex vivo MDSC differentiation methods with characteristics of tumor-infiltrating subsets would solve this problem. These methods should help to speed up the collection of information and the assessment of the activity of different therapeutics on human MDSCs.

Acknowledgments Grazyna Kochan research is funded by a CAIXA project grant from the, Spain, a Sandra Ibarra research grant, and a Gobierno Vasco BioEf project grant (BIO13/CI/014).

References

1. Youn JI, Kumar V, Collazo M, Nefedova Y, Condamine T, Cheng P, Villagra A, Antonia S, McCaffrey JC, Fishman M, Sarnaik A, Horna P, Sotomayor E, Gabrilovich DI (2013) Epigenetic silencing of retinoblastoma gene regulates pathologic differentiation of myeloid cells in cancer. Nat Immunol 14(3):211–220. doi:10.1038/ni.2526 ni.2526 [pii]
2. Liechtenstein T, Perez-Janices N, Gato M, Caliendo F, Kochan G, Blanco-Luquin I, Van der Jeught K, Arce F, Guerrero-Setas D, Fernandez-Irigoyen J, Santamaria E, Breckpot K, Escors D (2014) A highly efficient tumor-infiltrating MDSC differentiation system for

discovery of anti-neoplastic targets, which circumvents the need for tumor establishment in mice. Oncotarget 5(17):7843–7857

3. Bronte V, Wang M, Overwijk WW, Surman DR, Pericle F, Rosenberg SA, Restifo NP (1998) Apoptotic death of CD8+ T lymphocytes after immunization: induction of a suppressive population of Mac-1+/Gr-1+ cells. J Immunol 161(10):5313–5320

4. Kusmartsev S, Gabrilovich DI (2002) Immature myeloid cells and cancer-associated immune suppression. Cancer Immunol Immunother 51(6):293–298

5. Bronte V, Chappell DB, Apolloni E, Cabrelle A, Wang M, Hwu P, Restifo NP (1999) Unopposed production of granulocyte-macrophage colony-stimulating factor by tumors inhibits CD8+ T cell responses by dysregulating antigen-presenting cell maturation. J Immunol 162(10):5728–5737

6. Haile LA, Gamrekelashvili J, Manns MP, Korangy F, Greten TF (2010) CD49d is a new marker for distinct myeloid-derived suppressor cell subpopulations in mice. J Immunol 185 (1):203–210. doi:10.4049/jimmunol.0903573

7. Mandruzzato S, Solito S, Falisi E, Francescato S, Chiarion-Sileni V, Mocellin S, Zanon A, Rossi CR, Nitti D, Bronte V, Zanovello P (2009) IL4Ralpha+ myeloid-derived suppressor cell expansion in cancer patients. J Immunol 182(10):6562–6568. doi:10.4049/jimmunol.0803831

8. Rodriguez PC, Ernstoff MS, Hernandez C, Atkins M, Zabaleta J, Sierra R, Ochoa AC (2009) Arginase I-producing myeloid-derived suppressor cells in renal cell carcinoma are a subpopulation of activated granulocytes. Cancer Res 69(4):1553–1560. doi:10.1158/0008-5472.CAN-08-1921

9. Eruslanov E, Neuberger M, Daurkin I, Perrin GQ, Algood C, Dahm P, Rosser C, Vieweg J, Gilbert SM, Kusmartsev S (2012) Circulating and tumor-infiltrating myeloid cell subsets in patients with bladder cancer. Int J Cancer 130(5):1109–1119. doi:10.1002/ijc.26123

10. Gielen PR, Schulte BM, Kers-Rebel ED, Verrijp K, Petersen-Baltussen HM, ter Laan M, Wesseling P, Adema GJ (2015) Increase in both CD14-positive and CD15-positive myeloid-derived suppressor cell subpopulations in the blood of patients with glioma but predominance of CD15-positive myeloid-derived suppressor cells in glioma tissue. J Neuropathol Exp Neurol 74(5):390–400. doi:10.1097/NEN.0000000000000183

11. Vasquez-Dunddel D, Pan F, Zeng Q, Gorbounov M, Albesiano E, Fu J, Blosser RL, Tam AJ, Bruno T, Zhang H, Pardoll D, Kim Y (2013) STAT3 regulates arginase-I in myeloid-derived suppressor cells from cancer patients. J Clin Invest 123(4):1580–1589. doi:10.1172/JCI60083

12. Brandau S, Trellakis S, Bruderek K, Schmaltz D, Steller G, Elian M, Suttmann H, Schenck M, Welling J, Zabel P, Lang S (2011) Myeloid-derived suppressor cells in the peripheral blood of cancer patients contain a subset of immature neutrophils with impaired migratory properties. J Leukoc Biol 89(2):311–317. doi:10.1189/jlb.0310162

13. Filipazzi P, Valenti R, Huber V, Pilla L, Canese P, Iero M, Castelli C, Mariani L, Parmiani G, Rivoltini L (2007) Identification of a new subset of myeloid suppressor cells in peripheral blood of melanoma patients with modulation by a granulocyte-macrophage colony-stimulation factor-based antitumor vaccine. J Clin Oncol 25(18):2546–2553. doi:10.1200/JCO.2006.08. 5829

14. Tarhini AA, Butterfield LH, Shuai Y, Gooding WE, Kalinski P, Kirkwood JM (2012) Differing patterns of circulating regulatory T cells and myeloid-derived suppressor cells in metastatic melanoma patients receiving anti-CTLA4 antibody and interferon-alpha or TLR-9 agonist and GM-CSF with peptide vaccination. J Immunother 35(9):702–710. doi:10.1097/ CJI.0b013e318272569b

15. Kohanbash G, McKaveney K, Sakaki M, Ueda R, Mintz AH, Amankulor N, Fujita M, Ohlfest JR, Okada H (2013) GM-CSF promotes the immunosuppressive activity of glioma-infiltrating myeloid cells through interleukin-4 receptor-alpha. Cancer Res 73 (21):6413–6423. doi:10.1158/0008-5472.CAN-12-4124

16. Napolitano M, D'Alterio C, Cardone E, Trotta AM, Pecori B, Rega D, Pace U, Scala D, Scognamiglio G, Tatangelo F, Cacciapuoti C, Pacelli R, Delrio P, Scala S (2015) Peripheral myeloid-derived suppressor and T regulatory PD-1 positive cells predict response to neoadjuvant short-course radiotherapy in rectal cancer patients. Oncotarget 6(10):8261–8270

17. Wang L, Chang EW, Wong SC, Ong SM, Chong DQ, Ling KL (2013) Increased myeloid-derived suppressor cells in gastric cancer correlate with cancer stage and plasma S100A8/A9 proinflammatory proteins. J Immunol 190(2):794–804. doi:10.4049/jimmunol.1202088
18. Gros A, Turcotte S, Wunderlich JR, Ahmadzadeh M, Dudley ME, Rosenberg SA (2012) Myeloid cells obtained from the blood but not from the tumor can suppress T-cell proliferation in patients with melanoma. Clin Cancer Res 18(19):5212–5223. doi:10.1158/1078-0432.CCR-12-1108
19. Diaz-Montero CM, Salem ML, Nishimura MI, Garrett-Mayer E, Cole DJ, Montero AJ (2009) Increased circulating myeloid-derived suppressor cells correlate with clinical cancer stage, metastatic tumor burden, and doxorubicin-cyclophosphamide chemotherapy. Cancer Immunol Immunother 58(1):49–59. doi:10.1007/s00262-008-0523-4
20. Gehad AE, Lichtman MK, Schmults CD, Teague JE, Calarese AW, Jiang Y, Watanabe R, Clark RA (2012) Nitric oxide-producing myeloid-derived suppressor cells inhibit vascular E-selectin expression in human squamous cell carcinomas. J Invest Dermatol 132(11):2642–2651. doi:10.1038/jid.2012.190
21. Rodriguez PC, Ochoa AC (2008) Arginine regulation by myeloid derived suppressor cells and tolerance in cancer: mechanisms and therapeutic perspectives. Immunol Rev 222:180–191. doi:10.1111/j.1600-065X.2008.00608.x
22. Srivastava MK, Sinha P, Clements VK, Rodriguez P, Ostrand-Rosenberg S (2010) Myeloid-derived suppressor cells inhibit T-cell activation by depleting cystine and cysteine. Cancer Res 70(1):68–77
23. Corzo CA, Cotter MJ, Cheng P, Cheng F, Kusmartsev S, Sotomayor E, Padhya T, McCaffrey TV, McCaffrey JC, Gabrilovich DI (2009) Mechanism regulating reactive oxygen species in tumor-induced myeloid-derived suppressor cells. J Immunol 182(9):5693–5701
24. Yu J, Du W, Yan F, Wang Y, Li H, Cao S, Yu W, Shen C, Liu J, Ren X (2013) Myeloid-derived suppressor cells suppress antitumor immune responses through IDO expression and correlate with lymph node metastasis in patients with breast cancer. J Immunol 190(7):3783–3797. doi:10.4049/jimmunol.1201449 jimmunol.1201449 [pii]
25. Chevolet I, Speeckaert R, Schreuer M, Neyns B, Krysko O, Bachert C, Hennart B, Allorge D, van Geel N, Van Gele M, Brochez L (2015) Characterization of the immune network of IDO, tryptophan metabolism, PD-L1, and in circulating immune cells in melanoma. Oncoimmunology 4(3):e982382. doi:10.4161/2162402X.2014.982382
26. Novitskiy SV, Pickup MW, Gorska AE, Owens P, Chytil A, Aakre M, Wu H, Shyr Y, Moses HL (2011) TGF-beta receptor II loss promotes mammary carcinoma progression by Th17 dependent mechanisms. Cancer Discov 1(5):430–441. doi:10.1158/2159-8290.CD-11-0100
27. Yang L, Huang J, Ren X, Gorska AE, Chytil A, Aakre M, Carbone DP, Matrisian LM, Richmond A, Lin PC, Moses HL (2008) Abrogation of TGF beta signaling in mammary carcinomas recruits Gr-1+ CD11b+ myeloid cells that promote metastasis. Cancer Cell 13(1):23–35. doi:10.1016/j.ccr.2007.12.004
28. Condamine T, Ramachandran I, Youn JI, Gabrilovich DI (2015) Regulation of tumor metastasis by myeloid-derived suppressor cells. Ann Rev Med 66:97–110. doi:10.1146/annurev-med-051013-052304
29. Lechner MG, Liebertz DJ, Epstein AL (2010) Characterization of cytokine-induced myeloid-derived suppressor cells from normal human peripheral blood mononuclear cells. J Immunol 185(4):2273–2284. doi:10.4049/jimmunol.1000901 jimmunol.1000901 [pii]
30. Valenti R, Huber V, Filipazzi P, Pilla L, Sovena G, Villa A, Corbelli A, Fais S, Parmiani G, Rivoltini L (2006) Human tumor-released microvesicles promote the differentiation of myeloid cells with transforming growth factor-beta-mediated suppressive activity on T lymphocytes. Cancer Res 66(18):9290–9298
31. Gato-Cañas M, Martinez de Morentin X, Blanco-Luquin I, Fernandez-Irigoyen J, Zudaire I, Liechtenstein T, Arasanz H, Lozano T, Casares N, Knapp S, Chaikuad A, Guerrero-Setas D,

Escors D, Kochan G, Santamaria E (2015) A core of kinase-regulated interactomes defines the neoplastic MDSC lineage. Oncotarget (In press)

32. Boutte AM, McDonald WH, Shyr Y, Yang L, Lin PC (2011) Characterization of the MDSC proteome associated with metastatic murine mammary tumors using label-free mass spectrometry and shotgun proteomics. PLoS ONE 6(8):e22446. doi:10.1371/journal.pone. 0022446

33. Chornoguz O, Grmai L, Sinha P, Artemenko KA, Zubarev RA, Ostrand-Rosenberg S (2011) Proteomic pathway analysis reveals inflammation increases myeloid-derived suppressor cell resistance to apoptosis. Mol Cell Proteomics10(3):M110 002980. doi:10.1074/mcp.M110. 002980

Chapter 4
Ex Vivo MDSC Differentiation Models

David Escors and Grazyna Kochan

Abstract The development of adequate ex vivo cell differentiation models is the key for the study of cell ontology and functions. For example, since ex vivo differentiation systems for conventional myeloid dendritic cells (DCs) were developed, research in these important regulators of immunity was significantly increased. However, for other myeloid cell types such as myeloid-derived suppressor cells (MDSCs) this is more challenging. These cells are quite heterogeneous in phenotype and function, especially human MDSCs. Therefore, their mechanisms of differentiation are very poorly understood compared to other myeloid cell types. In recent years, several ex vivo differentiation methods have been developed to obtain cancer-specific MDSCs. Here we will describe some representative examples and briefly discuss their use and impact in MDSC research.

Keywords In vitro MDSC · GM-CSF · G-CSF · IL13 · IL6 · M-CSF · Macrophage · Tumor environment

4.1 Introduction

MDSCs differentiate from myeloid precursors present in the bone marrow of tumor-bearing patients. Growing tumors produce many cytokines and factors that distribute systemically through blood and lymphatics. These tumor-derived prod-

D. Escors (✉)
Navarrabiomed-Biomedical Research Centre, Fundación Miguel Servet, IdiSNA,
Calle Irunlarrea 3, 31008 Pamplona, Navarra, Spain
e-mail: descorsm@navarra.es

D. Escors
Immunomodulation Group Rayne Building, University College London, 5 University Street,
London WC1E 6JF, UK

G. Kochan
Navarrabiomed-Biomedical Research Centre, Fundación Miguel Servet, IdiSNA,
Irunlarrea 3, 31008 Pamplona, Navarra, Spain
e-mail: grkochan@navarra.es

© The Author(s) 2016
D. Escors et al., *Myeloid-Derived Suppressor Cells and Cancer*,
SpringerBriefs in Immunology, DOI 10.1007/978-3-319-26821-7_4

49

ucts reach the bone marrow, and disturb physiological myelopoiesis leading to fast mobilization of immature myeloid cells. These cells distribute systemically through the blood stream and accumulate at large numbers in immune organs and within the tumor itself. These myeloid cells possess strong immunosuppressive properties, particularly tumor-infiltrating subsets. There, MDSCs favor tumor progression and metastasis through the production of pro-angiogenic products and by inhibition of antitumor immune responses. These MDSCs also counteract chemo- and radio-therapy through the synthesis of ROS scavenging proteins and enzymes that degrade xenobiotics. Finally, as they acquire strong immunosuppressive properties, they inhibit antitumor immune cells including cytotoxic T cells, natural killer (NK), and dendritic cells (DCs).

The process of MDSC differentiation is thought to be quite complex and cancer-specific. The bone marrow microenvironment differs from that found in the spleen or within the tumor itself. Additionally, each tumor type secretes a different cytokine-chemokine profile. Thus, MDSC differentiation and the resulting MDSC subsets differ from cancer to cancer. Undoubtedly, intra-tumor MDSC subsets should be the target cells for research. However, their isolation is certainly a challenge, as low numbers of viable MDSCs are obtained from within tumors after a cumbersome isolation process. One way to circumvent this problem is to obtain MDSCs in vitro that model intra-tumor subsets. Therefore, appropriate ex vivo differentiation systems would eliminate the need of obtaining these MDSC subsets from in vivo. There are currently a "plethora" of different protocols and systems to generate MDSCs and MDSC-like cells, both in murine and human cancer systems. However, so far there is not a single protocol that has completely substituted MDSC subsets isolated from animals and human patients.

4.2   Ex Vivo Murine MDSC Differentiation Models

4.2.1   Differentiation Methods Based on Cancer Cell-Derived Conditioning Medium (Fig. 4.1a)

From early studies, it was apparent that medium from cultures of cancer cells contained cytokines that stimulated differentiation and expansion of suppressor cells from bone marrow precursors. This was clearly shown using supernatants from Lewis lung carcinoma cell cultures, which differentiated a monocyte-like population from bone marrow cells with immunosuppressive activities [1]. The authors concluded that the presence of a colony-stimulating factor secreted by these cells were responsible for differentiation of immunosuppressive cells.

Using a multiplex cytokine detection assay, it was demonstrated that a murine breast carcinoma cell line secreted high levels of GM-CSF, VEGF, and MCP-1. Out of these, the presence of GM-CSF was sufficient to induce MDSC differentiation from bone marrow haematopoietic precursors and keep these cells viable [2]. Interestingly, they showed that only the monocytic MDSC subset was suppressive.

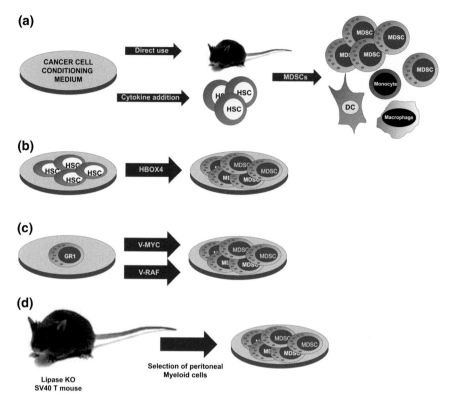

Fig. 4.1 Current methods for ex vivo MDSC differentiation. **a** In the scheme, the methods based on the use of conditioning medium obtained from cultures of cancer cells (lef). Precursors can be obtained either from bone marrow (usually murine) or peripheral blood (usually human). While MDSCs are obtained, differentiation cultures also contain a variable proportion of other contaminating myeloid cells types (right). HSC, haematopoietic stem cell; DC, dendritic cell. **b**, **c** Murine haematopoietic stem cells (HSC) or GR1+ myeloid cells can be transformed with retrovirus vectors expressing genes that favor myeloid differentiation (such as HBOX4) or immortalize cells (v-myc and v-raf), leading to homogeneous MDSC-like cultures. **d** Immortal MDSC-like cells have been isolated from lipase KO-SV40T mice, leading to homogenous cell lines, with strong suppressive activities

Nevertheless, neutralization of GM-CSF with specific blocking antibodies did not completely eliminate the capacity of the cancer-derived conditioning medium to differentiate MDSCs. However, the importance of GM-CSF in driving MDSC differentiation was demonstrated by the addition of recombinant GM-CSF to the breast cancer cell-derived medium, which improved cell viability.

Youn and cols demonstrated that MDSCs could be differentiated from mouse bone marrow by supplementing medium derived from different cancer cell types with GM-CSF and IL-4 [3, 4]. The authors observed a correlation between MDSC expansion in vivo with the capacity of the different conditioning media to differentiate MDSCs in vitro. Supernatants from cultures of EL4 lymphoma, B16 melanoma, and CT26 colorectal cancer cells were particularly efficient in driving MDSC differentiation in combination with recombinant GM-CSF and IL4. Nevertheless, the efficacy of this ex vivo method was limited, reaching not higher than 25% of CD11b+ GR1+ MDSCs, compared to 10% when using cell culture supernatants from control 3T3 mouse fibroblast cells. Importantly, using this ex vivo differentiation system and in vivo data, the authors described and characterized the two MDSC subsets for the first time in 10 different murine cancer models.

Most of the ex vivo protocols add recombinant myeloid-promoting cytokines to medium from cancer cell cultures. An interesting variation of this method relies on cytokine overexpression directly from cancer cells [5, 6]. These cytokine-expressing cancer cells produce conditioning medium which very efficiently drives MDSC differentiation from bone marrow precursors. These MDSC-like cells shared the phenotypic characteristics found in intra-tumor MDSCs, including specific iNOS upregulation and TGF-β secretion. A comparative analysis of their proteome with that of DC and noncancerous MDSC controls showed that these ex vivo MDSCs were metabolically active, and used lipid metabolism to obtain energy [6]. This metabolic profile was recently confirmed in MDSCs purified directly from tumors [7]. Using these ex vivo differentiation methods, it was confirmed in vitro that M-MDSCs are precursors of G-MDSCs, in agreement with in vivo observations [8].

Some of the previous studies indicated that apart from GM-CSF, other cytokines and molecules contribute to MDSC differentiation. Thus, Xiao and cols engineered murine prostate tumor cell lines to express the soluble version of the human NKG2D ligand MICB [9]. MIC expression on the surface of tumor cells contributes toward anti-tumor immunity by mediating NK and CD8 T-cell activation. However, tumor cells activate a protease that sheds a soluble version of MIC with strong immunosuppressive activities. The tumor cell conditioning medium expressing sMICB was supplemented with recombinant GM-CSF and used to differentiate MDSCs ex vivo from mouse bone marrow within 3 days. Interestingly, the authors achieved up to 70 % of differentiated MDSCs. This makes this system one of the most efficient described so far, together with cytokine overexpression directly from cancer cells.

De Veirman and cols demonstrated murine MDSC differentiation from BM-derived CD11b cells cultured in multiple myeloma 5T33MMvt cell-derived conditioning medium without GM-CSF supplement [10]. The authors showed that GM-CSF was already produced from murine myeloma cells. These ex vivo differentiated CD11b+ MDSCs efficiently inhibited the proliferation of anti-CD3/anti-CD28-activated T cells. Thus, it is undeniable that tumor cell-derived medium contains all the factors necessary for MDSC differentiation from myeloid precursors [11]. Xiang and cols demonstrated that those factors reside in tumor-derived exosomes, which contain PGE$_2$ and TGF-β. Neutralisation of these two molecules inhibits MDSC differentiation and abrogates the tumor-promoting effects of cancer cell-derived exosomes.

4.2.2 Differentiation Methods Based on Defined Media (Fig. 4.1a)

Using cancer cell-derived supernatants to induce MDSC differentiation is an experimental approach that tries to mimic the in vivo situation. However, tumors and cancer cells are heterogeneous, and it is likely that ex vivo differentiation systems that use cancer cell-derived supernatants may lack reproducibility, or achieve different outcomes depending on the cell lines or number of passages. Moreover, conditioning media obtained from cell cultures may significantly differ from batch to batch. Finally, the complex combination of cytokines and cell-derived products makes the study of the effects of single cytokines on MDSC differentiation a challenge. Therefore, the development of ex vivo MDSC differentiation systems based on defined cytokine combinations rather than culture supernatants should solve all these caveats. However, although MDSCs can be differentiated by adding certain recombinant cytokines to cell culture medium, the efficiency of differentiation is still rather poor [12].

Dendritic cells (DCs), the myeloid immunostimulatory counterparts, are easily differentiated ex vivo with recombinant GM-CSF. Interestingly, about 15 years ago it was observed that the simultaneous addition of lipopolysaccharide (LPS) and GM-CSF to mouse bone marrow cells resulted in the differentiation of a type of "immature DC"-like cell that exhibited T-cell immunosuppressive activities [13]. The authors described these cells as MHC$^{neg/low}$, CD14$^+$ and F4/80$^+$, and very likely corresponded to MDSCs. These cells were poor T-cell stimulators in mixed lymphocyte reactions (MLRs), and induced T-cell anergy, both classical features of MDSCs. In fact, it was later shown that these cells expanded from myeloid precursors generated in bone marrow-derived DC cultures especially with high GM-CSF concentrations. In addition, MDSC differentiation was shown to be enhanced by proinflammatory cytokines [13, 14].

Nevertheless, even though GM-CSF seems to be a very important cytokine for MDSC differentiation [15], the use of GM-CSF on its own is not sufficient to obtain fully competent MDSCs [16]. Marigo and cols tested GM-CSF, G-CSF, and IL6 either alone or in combination and found that either GM-CSF+G-CSF or GM-CSF + IL6 increased the efficiency of MDSC differentiation from mouse bone marrow compared to any of the above cytokines on their own [16]. These MDSCs were fully suppressive toward activated T cells and induced antigen unresponsiveness when inoculated in vivo. Furthermore, these cells could establish therapeutic tolerance in a mouse model for diabetes.

Highfill and cols used a combination of GM-CSF and G-CSF to obtain bone marrow-derived MDSCs. These MDSCs expressed both IL4R and F4/80 (a macrophage-specific marker). Ex vivo MDSCs were readily differentiated within 5 days, and the addition of IL13 significantly increased arginase 1 expression. Coadministration of MDSCs with T cells from C57BL/6 mice significantly inhibited graft-versus-host disease in Balb-c mice, without impairing the capacity of transferred T cells of mediating graft-versus-leukemia effects. Their strong regulatory activities correlated with arginase-1 expression, and the therapeutic effects could be replicated by direct administration of arginase-1. Thus, this study showed that ex vivo differentiated MDSCs can be utilized in cell-based therapies.

Apart from GM-CSF, production of proinflammatory cytokines by cancer cells was shown to drive MDSC differentiation from bone marrow precursors. Thus, IL6 in combination with GM-CSF leads to differentiation of an MDSC-like population with strong T-cell suppressive activities. MDSCs generated with this method by Hammami and cols showed a downregulation of central carbon metabolic pathway, including glycolysis, Kreb's cycle and glutaminolysis [17]. Importantly, these authors showed with this ex vivo system that MDSCs were metabolically active, causing a decrease in intracellular ATP concentration and consequently, AMPK activation. This AMPK activation was shown to regulate L-arginine metabolism and MDSC suppressive activities. In fact, AMPK was shown to be associated to tumor-infiltrating MDSCs, regulating glutathione metabolism [18].

Following the study by Xiao and cols mentioned in the previous section, the authors used purified sMICB in combination with recombinant GM-CSF to drive MDSC differentiation without the need of using tumor cell-derived conditioning medium [9]. Their method elicited a dose-dependent MDSC differentiation from bone marrow cells through STAT3 phosphorylation.

4.3   Ex Vivo Human MDSC Differentiation Models

Compared with murine MDSCs, much is unknown of human MDSCs. The ease of use of murine models facilitates MDSC research in this experimental system. Research in human MDSCs is still a challenge. Similarly to the mouse system, human bone marrow can be certainly used to isolate myeloid precursors and study in vitro MDSC differentiation. However, obtaining bone marrow from either healthy or cancer patients is an invasive procedure. Therefore, the use of haematopoietic precursors present in peripheral blood is preferred for human MDSC differentiation. Blood can be easily obtained in large quantities from human subjects without subjecting them to excessively invasive techniques. On the other hand, circulating haematopoietic precursors are present at low numbers in peripheral blood. For instance, mobilization of myeloid precursors to peripheral blood can be achieved by administration of G-CSF. However, this treatment may alter MDSC differentiation itself. According to published work, monocytes isolated from PBMCs are the preferred and most widely used source to obtain human MDSCs ex vivo. Even so, human MDSC differentiation is still less efficient compared to the murine system. And it is yet unclear whether ex vivo differentiated monocyte-derived MDSCs are a suitable working model for intra-tumor human MDSCs.

4.3.1 Differentiation Methods Based on Cancer Cell-Derived Conditioning Medium (Fig. 4.1a)

Myeloid conventional human DCs can be easily differentiated from PBMCs in the presence of recombinant GM-CSF and IL4. Thus, in analogy to the DC differentiation system, similar approaches were undertaken to derive MDSCs from human monocytes. Valenti and cols achieved MDSC differentiation from isolated monocytes following a protocol similar to DC differentiation, but adding tumor-derived microvesicles [19]. The resulting myeloid cells exhibited the "consensus" phenotype for human MDSCs (CD14+ HLA DR^{neg} and low expression of co-stimulatory molecules). These cells exerted dose-dependent T-cell suppressive activities which were mediated in part by TGF-β secretion. A similar result was obtained with microvesicles isolated from the plasma of melanoma patients, and MDSCs with similar phenotypic and functional characteristics were also shown to be circulating in peripheral blood from patients.

De Veirman and cols also demonstrated that MDSCs could be differentiated from human PBMCs in the presence of myeloma cell line-derived conditioning media [10]. Very interestingly, in this case GM-CSF was shown not to be expressed by human multiple myeloma cell lines, and alternative GM-CSF-independent mechanisms for MDSC differentiation were very likely implicated. The authors demonstrated that tumor cell-derived media contained factors that promoted MDSC survival, particularly Mcl-1.

4.3.2 Differentiation Methods Based on Defined Cytokine Combinations (Fig. 4.1a)

Lechner and cols quantified the expression of 15 different MDSC-inducing factors in a relatively large collection of tumor cell lines [20]. By separating cell lines known to induce MDSCs from those that do not, the authors were capable of identifying several molecules which could potentially drive MDSC differentiation. From these, COX2, IL1β, IL6, M-CSF, and IDO correlated with MDSC induction. Surprisingly, while Flt3L and SCF were expressed in by tumor cell lines, the expression of GM-CSF, TGFβ, and arginase 1 was downmodulated in MDSC-inducing tumor cell lines. Then, several cytokine combinations were tested to drive MDSC differentiation from PBMCs. Out of these, GMCSF+IL6 and GMCSF+IL6+VEGF yielded MDSCs with the strongest suppressive activities. GM-CSF alone, GM-CSF+IL1β, GM-CSF+TNF-α, and GM-CSF+VEGF combinations generated less potent immunosuppressive MDSCs. Interestingly, in the experimental conditions described in this study, MDSCs differentiated with GM-CSF and PGE$_2$ were weakly immunosuppressive while TGFβ seemed to interfere with MDSC-promoting cytokines. This study confirmed that in humans there are several distinct pathways leading to MDSC differentiation in cancer.

Similarly to the murine systems and the studies described above, Marigo and cols showed that human CD11b$^+$ CD16^{neg} MDSC-like cells could also be derived from human bone marrow precursors [16]. Bone marrow-derived myeloid cells using GM-CSF+G-CSF or GM-CSF+IL6 cytokine combinations showed strong T-cell inhibitory activities.

Obermajer and cols showed that human DC differentiation from monocytes was inhibited by the addition of PGE$_2$ to differentiation cultures using GM-CSF and IL4 [21]. Instead, PGE$_2$ induced the expression of COX2 in monocytes, shifting their differentiation toward MDSCs. It is well known that tumor cells overexpress PGE$_2$ and COX2, providing a mechanistical explanation for MDSC differentiation within the tumor environment. Ex vivo PGE$_2$-induced MDSCs from monocytes exhibited the expected phenotype for human MDSCs, which also corresponded to MDSCs isolated from human ovarian cancer ascites. These MDSCs also expressed COX2 and possessed strong T-cell suppressive activities.

The necessity of proinflammatory cytokines such as IL6 to drive human MDSC differentiation was clearly demonstrated for squamous cell carcinoma of the esophagus, both in a mouse model and also in human patients. Ex vivo MDSCs as defined by the CD11b$^+$ CD14$^+$ HLA DR^{neg} phenotype were derived from PBMCs by incubation with GM-CSF and IL6 [22]. Using this ex vivo system, IL6 was shown to stimulate hallmark immunosuppressive pathways in MDSCs, including STAT3 phosphorylation, ROS production, and arginase activity.

4.4   MDSC-Like Cell Lines

An attractive alternative to ex vivo MDSC differentiation is to develop established MDSC cell lines. Although it could be argued that MDSC cell lines may not fully reflect the in vivo complexity, these cell lines present many practical advantages. Large numbers of homogeneous MDSC-like cell preparations can be obtained in a short period of time. Ideally, these cell lines could be immortalized and grown indefinitely in vitro. These MDSC-like cell lines could be used for basic research and to screen anti-MDSC treatments. One of such strategies is the use of murine embryonic stem cells (ESCs) to derive MDSC-like cells [23]. This has been achieved by the overexpression of homeobox B4 in ESCs (Fig. 4.1b). In this way, myeloid differentiation was enhanced and MDSC-like cells were obtained using defined cytokine cocktails through three different culturing steps. Again, the manipulation of ESCs to enhance myeloid differentiation may not reflect a "physiological" situation, but the MDSC-like cells obtained in this fashion will surely be useful for cell therapy and drug screening.

Apolloni and cols generated immortalized murine spleen CD11b$^+$ Gr1$^+$ myeloid cells by retroviral expression of two oncogenes, v-myc and v-raf [24] (Fig. 4.1c). These MDSC-like cells possessed T-cell inhibitory activities. However, it has to be taken into account that spleen MDSC subsets are slightly different in phenotype and function from tumor-infiltrating subsets [25, 26]. In addition, the constitutive expression of two oncogenes will very likely perturb cell signaling pathways. Although for some studies this might not be important, this approach may interfere significantly with some applications such as screening of anticancer compounds.

Ding and cols employed a different approach to obtain cells modeling MDSCs. Thus, they obtained an immortal myeloid cell line that functionally resembled MDSCs. This cell line was isolated from peritoneal macrophages of mice generated by breading a SV40-T transgenic mouse with a lysosomal acid lipase knock-out strain [27] (Fig. 4.1c). These cells possessed strong T-cell immunosuppressive capacities, although it lacked expression of representative MDSC markers such as CD11b and Ly6G. The authors showed that this "MDSC" cell line had increased glucose metabolism. However, it has been shown that primary MDSCs strongly decrease glucose metabolism and precisely use lipid metabolism to generate the energy they need [6, 7]. Consequently, it is yet unclear whether these cells are truly MDSC-like or rather, macrophage-like.

4.5 Summary and Conclusions

The diversity of MDSC subsets and subtypes is a direct reflection of the in vivo high complexity. Growing tumors secrete high levels of cytokines and molecules that perturb haematopoiesis, leading to the mobilization of MDSCs with potent immunosuppressive and procarcinogenic properties. While physiological haematopoiesis is fairly well-known, the changes that take place during cancer progression are not well understood. This is also complicated by the fact that each tumor type will secrete cytokine profiles that are unique to that specific tumor. Attempts at replicating the tumor environment in vivo have shed some light in MDSC biology, particularly in murine experimental systems. Additionally, some specific and defined cytokine combinations drive MDSC differentiation from haematopoietic precursors. In contrast, human MDSCs are by far harder to study. Cancer patients are diagnosed and subjected to treatments that surely have an impact on MDSC types and numbers. Apart from this, obtaining bone marrow as a source of MDSC precursors subjects patients and donors to an invasive technique that does not yield high precursor numbers. Therefore, researchers have used monocytic cells isolated from peripheral blood as MDSC precursors. These protocols have been set up in analogy to DC differentiation methods that produce high DC numbers in vitro. However, these protocols are not as efficient in generating MDSCs. This is likely due to the lack of the presence of the "MDSC progenitor" in circulating peripheral blood. It is quite possible that this progenitor is different from DC precursors.

So, concluding, although major steps forward are being undertaken, an efficient protocol for human MDSC differentiation is still lacking. Once that human MDSC differentiation pathways have been defined, it is highly likely that very large MDSC numbers will be obtained in vivo for research and therapy.

Acknowledgments DE is funded by a Miguel Servet Fellowship (CP12/03114), a FIS project grant (PI14/00579) from the Instituto de Salud Carlos III, Spain, the Refbio transpyrenaic collaborative project grants (NTBM), a Sandra Ibarra Foundation grant, Gobierno de Navarra Grant (BMED 033-2014), and a Gobierno Vasco BioEf project grant (BIO13/CI/014). GK received a Caixa Bank research grant.

References

1. Young MR, Aquino S, Young ME (1989) Differential induction of hematopoiesis and immune suppressor cells in the bone marrow versus in the spleen by Lewis lung carcinoma variants. J Leukoc Biol 45(3):262–273
2. Morales JK, Kmieciak M, Knutson KL, Bear HD, Manjili MH (2010) GM-CSF is one of the main breast tumor-derived soluble factors involved in the differentiation of CD11b-Gr1- bone marrow progenitor cells into myeloid-derived suppressor cells. Breast Cancer Res Treat 123(1):39–49
3. Youn JI, Nagaraj S, Collazo M, Gabrilovich DI (2008) Subsets of myeloid-derived suppressor cells in tumor-bearing mice. J Immunol 181(8):5791–5802
4. Cheng P, Kumar V, Liu H, Youn JI, Fishman M, Sherman S, Gabrilovich D (2014) Effects of notch signaling on regulation of myeloid cell differentiation in cancer. Cancer Res 74(1): 141–152
5. Dufait I, Schwarze JK, Liechtenstein T, Leonard W, Jiang H, Law K, Verovski V, Escors D, De Ridder M, Breckpot K (2015) *Ex vivo* generation of myeloid-derived suppressor cells that model the tumor immunosuppressive environment in colorectal cancer Oncotarget 6 (14):12369–12382
6. Liechtenstein T, Perez-Janices N, Gato M, Caliendo F, Kochan G, Blanco-Luquin I, Van der Jeught K, Arce F, Guerrero-Setas D, Fernandez-Irigoyen J, Santamaria E, Breckpot K, Escors D (2014) A highly efficient tumor-infiltrating MDSC differentiation system for discovery of anti-neoplastic targets, which circumvents the need for tumor establishment in mice. Oncotarget 5(17):7843–7857
7. Hossain F, Al-Khami AA, Wyczechowska D, Hernandez C, Zheng L, Reiss K, Del Valle L, Trillo-Tinoco J, Maj T, Zou W, Rodriguez PC, Ochoa AC (2015) Inhibition of fatty acid oxidation modulates immunosuppressive functions of myeloid-derived suppressor cells and enhances cancer therapies. Cancer Immunol Res. doi:10.1158/2326-6066.CIR-15-0036
8. Youn JI, Kumar V, Collazo M, Nefedova Y, Condamine T, Cheng P, Villagra A, Antonia S, McCaffrey JC, Fishman M, Sarnaik A, Horna P, Sotomayor E, Gabrilovich DI (2013) Epigenetic silencing of retinoblastoma gene regulates pathologic differentiation of myeloid cells in cancer. Nat Immunol 14(3):211–220. doi:10.1038/ni.2526
9. Xiao G, Wang X, Sheng J, Lu S, Yu X, Wu JD (2015) Soluble NKG2D ligand promotes MDSC expansion and skews macrophage to the alternatively activated phenotype. Journal Hematol Oncol 8(1):13. doi:10.1186/s13045-015-0110-z
10. De Veirman K, Van Ginderachter JA, Lub S, De Beule N, Thielemans K, Bautmans I, Oyajobi BO, De Bruyne E, Menu E, Lemaire M, Van Riet I, Vanderkerken K, Van Valckenborgh E (2015) Multiple myeloma induces Mcl-1 expression and survival of myeloid-derived suppressor cells. Oncotarget 6(12):10532–10547
11. Xiang X, Poliakov A, Liu C, Liu Y, Deng ZB, Wang J, Cheng Z, Shah SV, Wang GJ, Zhang L, Grizzle WE, Mobley J, Zhang HG (2009) Induction of myeloid-derived suppressor cells by tumor exosomes. Int J Cancer 124(11):2621–2633. doi:10.1002/ijc.24249
12. Escors D, Liechtenstein T, Perez-Janices N, Schwarze J, Dufait I, Goyvaerts C, Lanna A, Arce F, Blanco-Luquin I, Kochan G, Guerrero-Setas D, Breckpot K (2013) Assessing T-cell responses in anticancer immunohterapy: dendritic cells or myeloid-derived suppressor cells? Oncoimmunology 12(10):e26148

13. Lutz MB, Kukutsch NA, Menges M, Rossner S, Schuler G (2000) Culture of bone marrow cells in GM-CSF plus high doses of lipopolysaccharide generates exclusively immature dendritic cells which induce alloantigen-specific CD4 T cell anergy in vitro. Eur J Immunol 30 (4):1048–1052

14. Rossner S, Voigtlander C, Wiethe C, Hanig J, Seifarth C, Lutz MB (2005) Myeloid dendritic cell precursors generated from bone marrow suppress T cell responses via cell contact and nitric oxide production in vitro. Eur J Immunol 35(12):3533–3544. doi:10.1002/eji.200526172

15. Bronte V, Chappell DB, Apolloni E, Cabrelle A, Wang M, Hwu P, Restifo NP (1999) Unopposed production of granulocyte-macrophage colony-stimulating factor by tumors inhibits CD8+ T cell responses by dysregulating antigen-presenting cell maturation. J Immunol 162(10):5728–5737

16. Marigo I, Bosio E, Solito S, Mesa C, Fernandez A, Dolcetti L, Ugel S, Sonda N, Bicciato S, Falisi E, Calabrese F, Basso G, Zanovello P, Cozzi E, Mandruzzato S, Bronte V (2010) Tumor-induced tolerance and immune suppression depend on the C/EBPbeta transcription factor. Immunity 32(6):790–802. doi:10.1016/j.immuni.2010.05.010 S1074-7613(10)00202-5 [pii]

17. Hammami I, Chen J, Murschel F, Bronte V, De Crescenzo G, Jolicoeur M (2012) Immunosuppressive activity enhances central carbon metabolism and bioenergetics in myeloid-derived suppressor cells in vitro models. BMC Cell Biol 13:18

18. Gato-Cañas M, Martinez de Morentin X, Blanco-Luquin I, Fernandez-Irigoyen J, Zudaire I, Liechtenstein T, Arasanz H, Lozano T, Casares N, Knapp S, Chaikuad A, Guerrero-Setas D, Escors D, Kochan G, Santamaria E (2015) A core of kinase-regulated interactomes defines the neoplastic MDSC lineage. Oncotarget In press

19. Valenti R, Huber V, Filipazzi P, Pilla L, Sovena G, Villa A, Corbelli A, Fais S, Parmiani G, Rivoltini L (2006) Human tumor-released microvesicles promote the differentiation of myeloid cells with transforming growth factor-beta-mediated suppressive activity on T lymphocytes. Cancer Res 66(18):9290–9298

20. Lechner MG, Liebertz DJ, Epstein AL (2010) Characterization of cytokine-induced myeloid-derived suppressor cells from normal human peripheral blood mononuclear cells. J Immunol 185(4):2273–2284. doi:10.4049/jimmunol.1000901 jimmunol.1000901 [pii]

21. Obermajer N, Muthuswamy R, Lesnock J, Edwards RP, Kalinski P (2013) Positive feedback between PGE2 and COX2 redirects the differentiation of human dendritic cells toward stable myeloid-derived suppressor cells. Blood 118(20):5498–5505

22. Chen MF, Kuan FC, Yen TC, Lu MS, Lin PY, Chung YH, Chen WC, Lee KD (2014) IL-6-stimulated CD11b+ CD14+ HLA-DR- myeloid-derived suppressor cells, are associated with progression and poor prognosis in squamous cell carcinoma of the esophagus. Oncotarget 5(18):8716–8728

23. Zhou Z, French DL, Ma G, Eisenstein S, Chen Y, Divino CM, Keller G, Chen SH, Pan PY (2010) Development and function of myeloid-derived suppressor cells generated from mouse embryonic and hematopoietic stem cells. Stem Cells (Dayton, Ohio) 28(3):620–632

24. Apolloni E, Bronte V, Mazzoni A, Serafini P, Cabrelle A, Segal DM, Young HA, Zanovello P (2000) Immortalized myeloid suppressor cells trigger apoptosis in antigen-activated T lymphocytes. J Immunol 165(12):6723–6730

25. Maenhout SK, Thielemans K, Aerts JL (2014) Location, location, location: functional and phenotypic heterogeneity between tumor-infiltrating and non-infiltrating myeloid-derived suppressor cells. Oncoimmunology 3(10):e956579. doi:10.4161/21624011.2014.956579

26. Maenhout SK, Van Lint S, Emeagi PU, Thielemans K, Aerts JL (2014) Enhanced suppressive capacity of tumor-infiltrating myeloid-derived suppressor cells compared to their peripheral counterparts. Int J Cancer 134(5):1077–1090. doi:10.1002/ijc.28449

27. Ding X, Wu L, Yan C, Du H (2015) Establishment of lal-/- myeloid lineage cell line that resembles myeloid-derived suppressive cells. PLoS ONE 10(3):e0121001. doi:10.1371/journal.pone.0121001

Chapter 5
Immunoregulatory Myeloid Cells in the Tumor Microenvironment

Jo A. Van Ginderachter

Abstract For decades, cancer therapies have been focused on attacking the cancer cells. However, in recent years it became clear that myeloid cells infiltrating the tumor are important players mediating tumor progression and metastasis. A prominent feature of tumor-associated myeloid cells is their immunoregulatory capacity, often leading to the subversion of antitumor T-cell immunity. Myeloid-derived suppressor cells (MDSC) are well known for their immunosuppressive capacity, while tumor-associated macrophages perform multiple functions including the skewing of T-cell responses. Tumor-associated dendritic cells consist of distinct populations with either pro- or antitumoral properties. A better understanding of the mechanisms that regulate the immunoregulatory capacity of myeloid cells in the tumor microenvironment will yield new avenues for therapeutic intervention that will be complementary to existing cancer cell-targeting approaches.

Keywords Macrophage · Dendritic cell · MDSC · Myeloid cell populations · Cancer · Tumor-associated macrophage · Tumor-associated dendritic cell

5.1 Introduction

Tumors comprise not only cancer cells, but also a heterogeneous group of non-cancerous cells such as endothelial cells, immune cells, and fibroblasts, as well as soluble factors and extracellular matrix. The tumor microenvironment regulates many aspects of cancer, from promoting neoplastic transformation, fostering therapeutic resistance, and protecting the tumor from host immunity leading to the promotion of tumor growth [1]. Notably, the tumor microenvironment is in constant

J.A. Van Ginderachter (✉)
VIB Lab Myeloid Cell Immunology, Building E, Vrije Universiteit Brussel,
Brussels, Belgium
e-mail: Jo.VanGinderachter@vib-vub.be

© The Author(s) 2016
D. Escors et al., *Myeloid-Derived Suppressor Cells and Cancer*,
SpringerBriefs in Immunology, DOI 10.1007/978-3-319-26821-7_5

evolution due to tissue remodeling, metabolic dysregulation, and the recruitment of various immune cell types to the tumor site.

The dominant populations of leukocytes in the tumor microenvironment are myeloid cells. Though these cells play a crucial and beneficial role as first line defenders against pathogens, they can be detrimental in cancer by supporting primary tumor growth and progression [2] and by enhancing metastasis through the induction of immature blood vessels and preparing the premetastatic niche [3]. Cancer cell–myeloid cell interactions are very complex, but these cells can use common pathways/mediators that lead to immune regulation and go hand in hand with angiogenesis [4]. This, together with emerging evidence on the plasticity of myeloid cell polarization opens the door to therapeutic strategies.

5.2 Myeloid Cells Within the Tumor Microenvironment and Their Immunoregulatory Role

Myeloid cells are the most abundant immune cells within the tumor and include at least four different populations.

(1) Tumor-associated macrophages (TAMs)

In solid tumors, TAM represents the dominant leukocyte population and their presence is associated with poor prognosis [5] due to (i) their promotion of angiogenesis and tissue remodeling via molecules such as VEGF, Bv8, and MMP9, and (ii) their inhibition of T-cell responses via the secretion of immunosuppressive cytokines such as IL-10 and TGF-β the L-arginine metabolism and the production of reactive oxygen species [6]. Importantly, plasticity is a hallmark of macrophages. On one hand, macrophages can acquire strong pro-inflammatory, tissue-destructive, antitumoral, and antimicrobial properties upon exposure to inflammatory cytokines and bacterial moieties (classically activated macrophages or M1). Conversely, alternatively activated macrophages or M2 are generally anti-inflammatory, promote tissue remodeling, repair, and angiogenesis upon exposure to a variety of triggers such as immunosuppressive or Th2 cytokines, glucocorticoids, or growth factors such as M-CSF [7]. The M1–M2 concept is obviously an oversimplification of the real-life situation. A spectrum model of human macrophage activation has been proposed [8] and novel nomenclature and experimental guidelines have been reported [9].

In a tumor context, chronic inflammation associated with the presence of M1 macrophages and the production of inflammatory mediators (TNFα, IL-6, ROS) may support neoplastic transformation [10]. At later stages of the disease, M2-type macrophages promote immune escape, tumor growth, and malignancy, and an M2 profile correlates with poor prognosis in several carcinomas [11, 12]. Most recently, differentially activated TAM subsets were reported to coexist in several transplantable mouse tumors residing in different tumor regions and performing distinct functions [13, 14]. These TAM subsets may include another specialized population,

Tie2-expressing macrophages (TEMs), which perform a nonredundant role in angiogenesis [15, 16].

The immunoregulatory role of macrophages in the tumor microenvironment

It has been demonstrated that the immune contexture of a tumor, i.e., the type, density, and location of immune cells in the tumor predicts the outcome [17]. In this context, the presence of CD8$^+$ cytotoxic T lymphocytes is often significantly correlated with an improved prognosis for the patient. As a countermeasure, tumors employ the immunoregulatory capacities of macrophages to suppress this acquired immunity and macrophages possess a plethora of mechanisms to do so.

A metabolic pathway well known to contribute to macrophage-mediated T-cell suppression is the L-arginine metabolism. L-arginine can be catabolized to the toxic moiety NO via the enzyme iNOS (inducible nitric oxide synthase or NOS2), or to polyamines and urea via arginase. High levels of aginase-1 are expressed by mature myeloid cells, presumably macrophages, in the tumor microenvironment of a mouse lung carcinoma model (Lewis lung carcinoma) [18]. Arginase-1 depletes L-arginine from the environment resulting in a downregulated CD3zeta expression in activated T cells. The importance of this pathway in vivo is demonstrated by a reduced tumor growth in mice treated with the arginase inhibitor N-hydroxy-nor-L-arginine. Notably, a high arginase expression level may be restricted to subsets of TAM, while iNOS activity may be more prominent in others. Employing MHC-II expression levels to discriminate between TAM subsets (M2-like MHC-II^{low} and M1-like MHC-II^{high} TAM), Movahedi et al. demonstrated an iNOS-dependent T-cell suppression by MHC-II^{high} TAM, while arginase-1 may be more relevant for the MHC-II^{low} cells [13]. Along the same line and employing the same model, MMR^{hi} TAM (which are likely similar to MHC-II^{low} TAM) were shown to express high levels of arginase-1 [19]. Interestingly, the Ron receptor tyrosine kinase was implicated in the arginase-1 expression through the induction of MAPK signaling, Fos activation, and binding to the AP-1 site in the *Arg* promoter. In addition, the intratumoral localization of TAM determines their immunosuppressive phenotype. MHC-II^{low} TAM are located in the most hypoxic areas [13] and prevent macrophages from entering the hypoxic tumor areas by blocking Neuropilin-1 function on these cells results in a reduced T-cell suppressive capacity [20]. Hence, while hypoxia does not govern monocyte-to-macrophage differentiation in tumors as such, it fine-tunes the protumoral functions of established TAM [21]. In accordance with these findings, HIF-1α was demonstrated to specifically regulate the immunosuppressive capacity of TAM, while not affecting their angiogenic capacity [22]. This is related to the high upregulation of arginase-1 under the combined effect of hypoxia-induced HIF-1α and soluble factors produced by mammary epithelial cells.

Arginase-1 is not the only immunosuppressive molecule that is induced by hypoxia. Also PD-L1, which shuts down T-cell activation by interacting with the immune checkpoint receptor PD-1, is upregulated under the influence of HIF-1α in macrophages and myeloid-derived suppressor cells (MDSC) [23]. Of note, PD-L1 can also be regulated on TAM via autocrine/paracrine IL-10 signaling, a mechanism which is shown to be responsible for glioma-induced immunosup-pression [24]. Likewise, autocrine IL-10, but also TNFα, were reported to induce PD-L1 expression on peritumoral stromal macrophages and monocytes in human hepatocellular carcinoma, which is crucial to shut down antitumor T-cell responses [25] or to induce protumoral Th22 cells [26]. Interestingly, the autocrine production of IL-10 and TNFα may be controlled by exposure of the macrophages/monocytes to IL-17 [27]. Moreover, it is important to realize that TAM also express CD80 (B7-1) and CD86 (B7-2), ligands for the T-cell inhibitory molecule CTLA-4, which may further contribute to a dampening of antitumor T-cell immunity. On the other hand, CD80 on tumor-associated phagocytes may also stimulate T-cell activity (upon interacting with CD28), but the expression of this costimulatory molecule can be inhibited by prostaglandin E_2 in breast tumors [28]. B7-H4 is another member of the B7 family of costimulatory molecules, for which the receptor is still elusive. B7-H4 is expressed on a subset of human ovarian carcinoma-associated TAM and contributes to the T-cell suppressive potential of these cells [29]. Finally, also HLA-G, a nonclassical MHC-I molecule that interacts with the inhibitory receptor ILT2 on T cells, can be upregulated on tumor-associated macrophages and released by these cells under the influence of neuroblastoma secreted factors [30].

TAM also secretes multiple factors that create an immunosuppressive network in the tumor microenvironment. For example, naturally occurring Treg is attracted to the tumor site under the influence of several chemokines that are mainly produced by TAM. In human ovarian carcinoma, TAM-secreted CCL22 recruits $CCR4^+$ Treg, thereby fostering tumor growth [31], while in colorectal carcinoma, CCL20 secreted by TAM recruits $CCR6^+$ Treg [32]. Also CCR5 can mediate Treg attraction to the tumor site and several CCR5 ligands (such as CCL5) were shown to be produced by TAM. Finally, CCL18 has been reported as a chemokine that is highly expressed by TAM from human breast tumors [33]. CCL18 has the capacity to recruit Treg to the lung [34], but its involvement in Treg recruitment to tumors awaits further experimentation.

Typical immunosuppressive cytokines, such as IL-10 and TGF-β, are also produced by TAM. These cytokines have the capacity to induce Treg de novo. In addition, IL-10 has a major impact on adaptive immunity by functioning as a negative regulator of Th1 and Th2 immunity [35], and so does TGF-β [36].

(2) Myeloid-derived suppressor cells (MDSC)

MDSC are a heterogeneous myeloid cell population whose common characteristics are an immature state and their ability to suppress T-cell responses. They have been abundantly observed in cancer, both in mice and in humans, and they accumulate within primary and metastatic tumors, bone marrow, spleen, and blood [37, 38]. Two main MDSC subpopulations have been characterized in mice—monocytic MO-MDSC ($CD11b^+Ly6G^-Ly6C^{hi}$) and granulocytic PMN-MDSC ($CD11b^+Ly6G^+Ly6C^{low}$) [39, 40]. Their equivalents in humans have been described, and are recognized as $Lin^-CD11b^+MHC-II^{lo}CD14^+$ MO-MDSC and $Lin^-CD11b^+MHC-II^{lo}CD15^+$ PMN-MDSC [41]. These two populations depend on different factors for their expansion/survival—MO- but not PMN-MDSC are expanded by GM-CSF [42, 43]—

and exert their suppressive function via different mechanisms. MO-MDSC have been reported to be more suppressive on a per cell basis [39, 42], mainly via iNOS [39] and in a contact-dependent but nonantigen-specific manner [44]. In contrast, PMN-MDSC suppress antigen-specific responses in a ROS-dependent manner [45]. Furthermore, their distribution in tumors and periphery is different: while PMN-MDSC are most abundant in blood, spleen, and bone marrow, MO-MDSC are more abundant within the majority of tumors [40]. The reasons for this may be due to differential recruitment or intratumoral expansion.

The immunoregulatory role of MDSC in the tumor microenvironment

The vast majority of studies on the immunoregulatory functions of MDSC have been performed with cells isolated from the spleen. Though secondary lymphoid organs are typically the sites of antitumor immunity induction, activated T cells may also undergo suppression at the tumor site. Findings with splenic MDSC may not be easily translatable to tumor-associated MDSC, since it has been demonstrated that the suppressive capacity of MDSC is heavily influenced by their location. In models of acute prostate inflammation and prostate cancer, only the MDSC derived from the inflammatory or tumor site (and not from the spleen) possess an immediate T-cell suppressive capacity, linked with their high expression of arginase-1, iNOS, and the MDSC-inducing transcription factor c-EBPβ [46]. These authors, hence suggested that an efficient suppression of T cells is restricted to the site of inflammation and is absent in the periphery. Along the same line, MDSC isolated from mouse lymphoma and melanoma tumors are more suppressive on a per cell basis than their splenic counterparts from the same mice, which was shown by two independent studies [47, 48]. Besides an increased NO production and arginase activity, the tumor-derived cells also expressed higher levels of CD80, which seem to be mediated via cell contact with the cancer cells. Remarkably, $CD80^+$ MDSC are more suppressive than $CD80^-$ cells, suggestive of a CD80-CTLA-4 interaction as mediator of suppression [48]. Moreover, CD80 can also bind to PD-L1, which results in the delivery of inhibitory signals to T cells. The upregulation of CD80 on MDSC was also seen in a mouse model of ovarian carcinoma, which again depended on an interaction of the MDSC with the cancer cells [49]. In this model, CD80 seems to activate Treg through its interaction with CTLA-4, and an antibody-mediated blockade of either CD80 or CTLA-4 alleviates suppression. Notably, though arginase-1 and iNOS are higher in tumor-associated MDSC, ROS production is elevated in the splenic counterparts [47]. Hence, the tumor microenvironment drives the gene expression of *Arg1* and *Nos2*, while lowering the expression of $p47^{phox}$ and $gp91^{phox}$. In addition, MDSC reaching the tumor site rapidly acquire a propensity to differentiate into macrophages.

An important microenvironmental cue that drives the phenotype of MDSC in tumors is hypoxia. Culturing splenic MDSC under hypoxic conditions recapitulates the characteristics of tumor-associated MDSC, including the expression levels of *Arg1*, *Nos2,* and the NADPH oxidase complex [47]. The hypoxia inducible transcription factor HIF-1α appears instrumental, since MDSC with a conditional HIF-1α deficiency fail to adopt the tumor-associated phenotype, do not differentiate into macrophages and lose viability more rapidly. Recently, PD-L1 was shown to be induced in MDSC by hypoxia due to the presence of a hypoxia responsive element (HRE) in the gene's proximal promoter region [23]. PD-L1 is at least partly involved in the suppressive capacity of hypoxic MDSC as blocking this molecule reduces T-cell suppression upon polyclonal or antigen-specific stimulation. Remarkably, the suppressive role of PD-L1 is not direct, but rather due to its indirect effect on enhancing IL-6 and IL-10 secretion by hypoxic MDSC. IL-10 is ultimately the effector molecule, since anti-IL-10 antibodies also lower the suppressive capacity of hypoxic MDSC [23]. How PD-L1 ligation is linked to the regulation of the *il10* gene is currently unclear. Of note, IL-10 has also been shown to be highly produced by MDSC and to be one of the main drivers of MDSC's suppressive capacity in a model of ovarian carcinoma [50].

(3) **Tumor-associated Dendritic cells (TADC)**

Dendritic cells are differentiated myeloid cells that specialize in antigen processing and presentation to naïve T cells. Human and mouse CD11c$^+$ DC subsets can be organized into four broad subsets—based on shared phenotypic markers and functional specialization—irrespective of their primary location in secondary lymphoid organs or in the parenchyma of nonlymphoid organs: (i) CD8α$^+$ DC-like cells (cDC1), (ii) CD11b$^+$ DC-like cells (cDC2) (iii) CD11b$^+$Ly6C$^+$ monocyte-derived DC and (iv) SiglecH$^+$ plasmacytoid DC (pDC) [51]. The main characteristic of DC is their ability to mature in response to stimuli such as pathogen- or danger-associated molecular patterns. Like macrophages, classically activated DCs, through the upregulation of costimulatory molecules and cytokines such as IL-12, are immunogenic, while "alternatively" activated or semi-mature DCs induce T-cell tolerance via deletion, anergy, or induction of regulatory T cells [52]. This again is an oversimplified view and it is likely that TADC exist in a multitude of functional states, and may be conditioned by the tumor to maintain immune tolerance or suppression [53].

The immunoregulatory role of DC in the tumor microenvironment

The infiltration of tumors by DC has been associated with good prognosis in several cancers [54], in particular at early stages [55], suggesting an active involvement of tumor-associated DC (TADC) in initiating antitumor adaptive immunity. Nevertheless, numerous defects in TADC have been reported as well, which may indicate that the pro- versus antitumoral role of TADCs depends on the stage of tumor growth. Indeed, in a spontaneous model of ovarian carcinoma development, immunostimulatory DC were present in the early stages of tumor growth—when the tumor was still kept under control by the immune system (equilibrium phase)—while immunosuppressive TADC were induced in the progressive phase of the tumors [56]. Tumor-derived prostaglandin E$_2$ and TGF-β1 were involved in the promotion of this immunosuppressive phenotype.

Even during later stages of tumor growth, a minor population of TADC could still be responsible for the initiation of antitumor T cells. CD8α$^+$ cDC (also known as cDC1) are not very abundant in a spontaneous mouse model of breast carcinoma, but are unique in their antigen processing and presenting capacity and their ability to stimulate naïve and activated T cells [57]. These rare cells contribute to T-cell interactions in tumors and compete for T-cell occupancy when located near the tumor margins. In addition, their presence is crucial for adoptive T-cell therapy. The presence and maturation of these cDC1 cells may be dependent on type I interferon. Indeed, IFNAR1 (IFNα/β receptor 1)-deficient $CD8\alpha^+$ TADC display defects in their cross-priming capacity to $CD8^+$ T cells [58]. Remarkably, IFN-β is produced by the TADC themselves, illustrating the presence of an autocrine activation loop [59]. In a melanoma model with a cancer cell-intrinsic β-catenin signaling, the transcriptional repressor ATF3 is induced, which results in a strongly diminished production of the chemokine CCL4 that is responsible for the attraction of immunostimulatory $CD8\alpha^+$ cDC [60].

As indicated above, TADC in bulkier tumors are often dysfunctional and multiple mechanisms have been reported. The lipid metabolism inside tumors may be one driver of TADC skewing. Human and mouse tumors produce ligands for the nuclear hormone receptor LXR that inhibit CCR7 expression on maturing DCs, and therefore their migration to secondary lymphoid organs. Preventing the cancer cells from producing these lipid LXR ligands result in the induction of an effective antitumor T-cell response [61]. Also other extracellular lipids are taken up by TADC via the scavenger receptor A [62]. Accumulation of these lipids in the DC caused a major defect in their antigen processing capacity and hence, their ability to stimulate T cells. More specifically, oxidized lipids (including triglycerides, cholesterol esters, and fatty acids), but not nonoxidized lipids, blocked cross-presentation by TADC by reducing the expression of peptide-MHC class I complexes at the cell surface [63]. Remarkably, besides lipid uptake, intracellular triglyceride biosynthesis due to the activation of the ER stress response factor XBP1 results in TADC lipid accumulation and a reduced support for antitumor T cells [64]. ER stress in TADC appears to be the result of lipid peroxidation byproducts. Finally, a reduced ability to present antigens can also be mediated by an altered metabolism that involves the interaction between pyruvate kinase M2 and SOCS3, leading to a reduced ATP production [65].

Besides hampering antigen presentation, TADC may also play a protumoral role by inducing tumor-specific T-cell tolerance via the upregulation of inhibitory molecules such as B7-H1 [66], or by the production of arginase [67], oxygen-dependent pathways that downregulate CD3 [68] or IDO [69].

5.3 Concluding Remarks

Cross talk between cancer cells and myeloid cells is complex and dynamic. However, common mechanisms, cellular players, and factors underly angiogenesis and immune suppression, thereby opening the door to therapeutic intervention. Further efforts are needed to fully understand the plasticity of tumor-associated myeloid cells, not only in terms of their activation state, but also in terms of their differentiation and interconversion. This emerging field holds the promise of identifying novel strategies aimed at manipulating the phenotype of these tumor-promoting cells.

References

1. Swartz MA, Iida N, Roberts EW et al (2012) Tumor microenvironment complexity: emerging roles in cancer therapy. Cancer Res 72(10):2473–2480
2. Mantovani A, Sica A (2010) Macrophages, innate immunity and cancer: balance, tolerance, and diversity. Curr Opin Immunol 22(2):231–237
3. Peinado H, Lavotshkin S, Lyden D (2011) The secreted factors responsible for pre-metastatic niche formation: old sayings and new thoughts. Semin Cancer Biol 21(2):139–146
4. Motz GT, Coukos G (2011) The parallel lives of angiogenesis and immunosuppression: cancer and other tales. Nat Rev Immunol 11(10):702–711
5. Bingle L, Brown NJ, Lewis CE (2002) The role of tumour-associated macrophages in tumour progression: implications for new anticancer therapies. J Pathol 196(3):254–265
6. Laoui D, Van Overmeire E, Movahedi K et al (2011) Mononuclear phagocyte heterogeneity in cancer: different subsets and activation states reaching out at the tumor site. Immunobiology 216(11):1192–1202
7. Murray PJ, Wynn TA (2011) Protective and pathogenic functions of macrophage subsets. Nat Rev Immunol 11(11):723–737
8. Xue J, Schmidt SV, Sander J et al (2014) Transcriptome-based network analysis reveals a spectrum model of human macrophage activation. Immunity 40(2):274–288
9. Murray PJ, Allen JE, Biswas SK et al (2014) Macrophage activation and polarization: nomenclature and experimental guidelines. Immunity 41(1):14–20
10. Sica A, Bronte V (2007) Altered macrophage differentiation and immune dysfunction in tumor development. J Clin Invest 117(5):1155–1166
11. Galon J, Costes A, Sanchez-Cabo F et al (2006) Type, density, and location of immune cells within human colorectal tumors predict clinical outcome. Science 313(5795):1960–1964
12. Kurahara H, Shinchi H, Mataki Y et al (2011) Significance of M2-polarized tumor-associated macrophage in pancreatic cancer. J Surg Res 167(2):e211–e219
13. Movahedi K, Laoui D, Gysemans C et al (2010) Different tumor microenvironments contain functionally distinct subsets of macrophages derived from Ly6C(high) monocytes. Cancer Res 70(14):5728–5739
14. Movahedi K, Schoonooghe S, Laoui D et al (2012) Nanobody-based targeting of the Macrophage Mannose Receptor for effective in vivo imaging of tumor-associated macrophages. Cancer Res 19 (E-pub ahead of print) 72(16):4165–4177
15. Coffelt SB, Tal AO, Scholz A et al (2010) Angiopoietin-2 regulates gene expression in TIE2-expressing monocytes and augments their inherent proangiogenic functions. Cancer Res 70(13):5270–5280
16. De Palma M, Venneri MA, Galli R et al (2005) Tie2 identifies a hematopoietic lineage of proangiogenic monocytes required for tumor vessel formation and a mesenchymal population of pericyte progenitors. Cancer Cell 8(3):211–226
17. Galon J, Costes A, Sanchez-Cabo F et al (2006) Type, density, and location of immune cells within human colorectal tumors predict clinical outcome. Science 313(5795):1960–1964
18. Rodriguez PC, Quiceno DG, Zabaleta J et al (2004) Arginase I production in the tumor microenvironment by mature myeloid cells inhibits T-cell receptor expression and antigen-specific T-cell responses. Cancer Res 64(16):5839–5849
19. Sharda DR, Yu S, Ray M et al (2011) Regulation of macrophage arginase expression and tumor growth by the Ron receptor tyrosine kinase. J Immunol 187(5):2181–2192

20. Casazza A, Laoui D, Wenes M et al (2013) Impeding macrophage entry into hypoxic tumor areas by Sema3A/Nrp1 signaling blockade inhibits angiogenesis and restores antitumor immunity. Cancer Cell 24(6):695–709
21. Laoui D, Van Overmeire E, Di Conza G et al (2014) Tumor hypoxia does not drive differentiation of tumor-associated macrophages but rather fine-tunes the M2-like macrophage population. Cancer Res 74(1):24–30
22. Doedens AL, Stockmann C, Rubinstein MP et al (2010) Macrophage expression of hypoxia-inducible factor-1 alpha suppresses T-cell function and promotes tumor progression. Cancer Res 70(19):7465–7475
23. Noman MZ, Desantis G, Janji B et al (2014) PD-L1 is a novel direct target of HIF-1α, and its blockade under hypoxia enhanced MDSC-mediated T cell activation. J Exp Med 211(5): 781–790
24. Bloch O, Crane CA, Kaur R, Safaee M, Rutkowski MJ, Parsa AT (2013) Gliomas promote immunosuppression through induction of B7-H1 expression in tumor-associated macrophages. Clin Cancer Res 19(12):3165–3175
25. Kuang DM, Zhao Q, Peng C, Xu J, Zhang JP, Wu C, Zheng L (2009) Activated monocytes in peritumoral stroma of hepatocellular carcinoma foster immune privilege and disease progression through PD-L1. J Exp Med 206(6):1327–1337
26. Kuang DM, Xiao X, Zhao Q et al (2014) B7-H1-expressing antigen-presenting cells mediate polarization of protumorigenic Th22 subsets. J Clin Invest 124(10):4657–4667
27. Zhao Q, Xiao X, Wu Y, Wei Y, Zhu LY, Zhou J, Kuang DM (2011) Interleukin-17-educated monocytes suppress cytotoxic T-cell function through B7-H1 in hepatocellular carcinoma patients. Eur J Immunol 41(8):2314–2322
28. Olesch C, Sha W, Angioni C et al (2015) MPGES-1-derived PGE2 suppresses CD80 expression on tumor-associated phagocytes to inhibit anti-tumor immune responses in breast cancer. Oncotarget 6(12):10284–10296
29. Kryczek I, Zou L, Rodriguez P et al (2006) B7-H4 expression identifies a novel suppressive macrophage population in human ovarian carcinoma. J Exp Med 203(4):871–881
30. Morandi F, Levreri I, Bocca P, Galleni B, Raffaghello L, Ferrone S, Prigione I, Pistoia V (2007) Human neuroblastoma cells trigger an immunosuppressive program in monocytes by stimulating soluble HLA-G release. Cancer Res 67(13):6433–6441
31. Curiel TJ, Coukos G, Zou L et al (2004) Specific recruitment of regulatory T cells in ovarian carcinoma fosters immune privilege and predicts reduced survival. Nat Med 10(9):942–949
32. Liu J, Zhang N, Li Q et al (2011) Tumor-associated macrophages recruit CCR6+ regulatory T cells and promote the development of colorectal cancer via enhancing CCL20 production in mice. PLoS ONE 6(4):e19495
33. Chen J, Yao Y, Gong C et al (2011) CCL18 from tumor-associated macrophages promotes breast cancer metastasis via PITPNM3. Cancer Cell 19(4):541–555
34. Chenivesse C, Chang Y, Azzaoui I et al (2012) Pulmonary CCL18 recruits human regulatory T cells. J Immunol 189(1):128–137
35. Ng TH, Britton GJ, Hill EV, Verhagen J, Burton BR, Wraith DC (2013) Regulation of adaptive immunity; the role of interleukin-10. Front Immunol 4:129
36. Oh SA, Li MO (2013) TGF-b: guardian of T cell function. J Immunol 191(8):3973–3979
37. Gabrilovich DI, Nagaraj S (2009) Myeloid-derived suppressor cells as regulators of the immune system. Nat Rev Immunol 9(3):162–174
38. Ostrand-Rosenberg S, Sinha P (2009) Myeloid-derived suppressor cells: linking inflammation and cancer. J Immunol 182(8):4499–4506
39. Movahedi K, Guilliams M, Van den Bossche J et al (2008) Identification of discrete tumor-induced myeloid-derived suppressor cell subpopulations with distinct T cell-suppressive activity. Blood 111(8):4233–4244
40. Youn J-I, Nagaraj S, Collazo M, Gabrilovich DI (2008) Subsets of myeloid-derived suppressor cells in tumor-bearing mice. J Immunol 181(8):5791–5802

41. Filipazzi P, Huber V, Rivoltini L (2012) Phenotype, function and clinical implications of myeloid-derived suppressor cells in cancer patients. Cancer Immunol Immunother 61(2): 255–263
42. Dolcetti L, Peranzoni E, Ugel S et al (2010) Hierarchy of immunosuppressive strength among myeloid-derived suppressor cell subsets is determined by GM-CSF. Eur J Immunol 40(1):22–35
43. Lesokhin AM, Hohl TM, Kitano S et al (2012) Monocytic CCR2+ myeloid-derived suppressor cells promote immune escape by limiting activated CD8 T-cell infiltration into the tumor microenvironment. Cancer Res 72(4):876–886
44. Solito S, Bronte V, Mandruzzato S (2011) Antigen specificity of immune suppression by myeloid-derived suppressor cells. J Leukoc Biol 90(1):31–36
45. Nagaraj S, Gupta K, Pisarev V et al (2007) Altered recognition of antigen is a mechanism of CD8+ T cell tolerance in cancer. Nat Med 13(7):828–835
46. Haverkamp JM, Crist SA, Elzey BD, Cimen C, Ratliff TL (2011) In vivo suppressive function of myeloid-derived suppressor cells is limited to the inflammatory site. Eur J Immunol 41 (3):749–759
47. Corzo CA, Condamine T, Lu L et al (2010) HIF-1α regulates function and differentiation of myeloid-derived suppressor cells in the tumor microenvironment. J Exp Med 207(11): 2439–2453
48. Maenhout SK, Van Lint S, Emeagi PU, Thielemans K, Aerts JL (2014) Enhanced suppressive capacity of tumor-infiltrating myeloid-derived suppressor cells compared with their peripheral counterparts. Int J Cancer 134(5):1077–1090
49. Yang R, Cai Z, Zhang Y, Yutzy WH 4th, Roby KF, Roden RB (2006) CD80 in immune suppression by mouse ovarian carcinoma-associated Gr-1+ CD11b+ myeloid cells. Cancer Res 66(13):6807–6815
50. Hart KM, Byrne KT, Molloy MJ, Usherwood EM, Berwin B (2011) IL-10 immunomodulation of myeloid cells regulates a murine model of ovarian cancer. Front Immunol 2:29
51. Guilliams M, Henri S, Tamoutounour S et al (2010) From skin dendritic cells to a simplified classification of human and mouse dendritic cell subsets. Eur J Immunol 40(8):2089–2094
52. Joffre O, Nolte MA, Spörri R, e Sousa CR (2009) Inflammatory signals in dendritic cell activation and the induction of adaptive immunity. Immunol Rev 227(1):234–247
53. Ma Y, Aymeric L, Locher C, Kroemer G, Zitvogel L (2011) The dendritic cell-tumor cross-talk in cancer. Curr Opin Immunol 23(1):146–152
54. Fridman W-H, Galon J, Pagès F et al (2011) Prognostic and predictive impact of intra- and peritumoral immune infiltrates. Cancer Res 71(17):5601–5605
55. Hiraoka N, Yamazaki-Itoh R, Ino Y et al (2011) CXCL17 and ICAM2 are associated with a potential anti-tumor immune response in early intraepithelial stages of human pancreatic carcinogenesis. Gastroenterology 140(1):310–321
56. Scarlett UK, Rutkowski MR, Rauwerdink AM et al (2012) Ovarian cancer progression is controlled by phenotypic changes in dendritic cells. J Exp Med 209(3):495–506
57. Broz ML, Binnewies M, Boldajipour B et al (2014) Dissecting the tumor myeloid compartment reveals rare activating antigen-presenting cells critical for T cell immunity. Cancer Cell 26(5):638–652
58. Diamond MS, Kinder M, Matsushita H et al (2011) Type I interferon is selectively required by dendritic cells for immune rejection of tumors. J Exp Med 208(10):1989–2003
59. Fuertes MB, Kacha AK, Kline J, Woo SR, Kranz DM, Murphy KM, Gajewski TF (2011) Host type I IFN signals are required for antitumor CD8+ T cell responses through CD8α+ dendritic cells. J Exp Med 208(10):2005–2016
60. Spranger S, Bao R, Gajewski TF (2015) Melanoma-intrinsic β-catenin signalling prevents anti-tumour immunity. Nature 523(7559):231–235
61. Villablanca EJ, Raccosta L, Zhou D et al (2010) Tumor-mediated liver X receptor-alpha activation inhibits CC chemokine receptor-7 expression on dendritic cells and dampens antitumor responses. Nat Med 16(1):98–105

62. Herber DL, Cao W, Nefedova Y et al (2010) Lipid accumulation and dendritic cell dysfunction in cancer. Nat Med 16(8):880–886
63. Ramakrishnan R, Tyurin VA, Veglia F et al (2014) Oxidized lipids block antigen cross-presentation by dendritic cells in cancer. J Immunol 192(6):2920–2931
64. Cubillos-Ruiz JR, Silberman PC, Rutkowski MR et al (2015) ER stress sensor XBP1 controls anti-tumor immunity by disrupting dendritic cell homeostasis. Cell 161(7):1527–1538
65. Zhang Z, Liu Q, Che Y et al (2010) Antigen presentation by dendritic cells in tumors is disrupted by altered metabolism that involves pyruvate kinase M2 and its interaction with SOCS3. Cancer Res 70(1):89
66. Curiel TJ, Wei S, Dong H et al (2003) Blockade of B7-H1 improves myeloid dendritic cell-mediated antitumor immunity. Nat Med 9(5):562–567
67. Liu Q, Zhang C, Sun A et al (2009) Tumor-educated CD11bhighIalow regulatory dendritic cells suppress T cell response through arginase I. J Immunol 182(10):6207–6216
68. Kuang D-M, Zhao Q, Xu J et al (2008) Tumor-educated tolerogenic dendritic cells induce CD3epsilon down-regulation and apoptosis of T cells through oxygen-dependent pathways. J Immunol 181(5):3089–3098
69. Muller AJ, Sharma MD, Chandler PR et al (2008) Chronic inflammation that facilitates tumor progression creates local immune suppression by inducing indoleamine 2,3 dioxygenase. Proc Natl Acad Sci USA 105(44):17073–17078

Chapter 6
Signal Transducer and Activation of Transcription 3: A Master Regulator of Myeloid-Derived Suppressor Cells

Karine Breckpot

Abstract Cancer progression is determined by cancer cells as well as various immune cells that make up the tumor microenvironment (TME). These immune cells consist of so-called effector cells such as natural killer cells and cytotoxic T lymphocytes, which fight cancer progression, and of immunosuppressive immune cells including regulatory T cells and immature myeloid cells, which aid tumor progression. Immature myeloid cells in the TME are further divided into several populations, amongst which are tumor-associated dendritic cells (TADC), tumor-associated macrophages (TAM), and myeloid-derived suppressor cells (MDSC). While TADC and TAM can be found in different activation states, MDSC can be further subdivided in two subsets: polymorphonuclear and monocytic MDSC. In recent years, MDSC received much attention as they are believed to exert a plethora of inhibitory mechanisms to create a tumor promoting TME. The recruitment, activation, and function of MDSC in the TME are largely determined by the transcription factor signal transducer and activator of transcription 3 (STAT3). Therefore, this review will focus on the role of this key signaling pathway during the MDSC life cycle.

Keywords STAT3 · Growth factors · Tumor microenvironment · Macrophages · Tregs · Phosphorylation

6.1 An Introduction to Tumor Immunology and Myeloid-Derived Suppressor Cells

The immune system can discriminate cancer cells from the healthy cells they originate from based on the expression of tumor antigens [1]. Tumor antigens can be ingested by antigen-presenting cells (APC) such as dendritic cells (DC) that infiltrate the tumor. While these APC migrate towards tumor-draining lymph nodes,

K. Breckpot (✉)
Faculty of Medicine and Pharmacy, Vrije Universiteit Brussel, Brussel, Belgium
e-mail: Karine.Breckpot@vub.ac.be

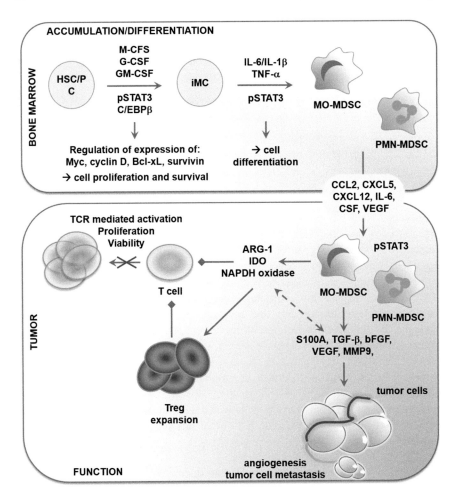

Fig. 6.1 Role of STAT3 in accumulation, differentiation, and functional regulation of MDSC in cancer. Cytokines such as M-, G- and GM-CSF stimulate myeloid cell development from HSC. Increased production of these cytokines during tumorogenesis interferes with normal myeloid development resulting in the generation of iMC. In presence of factors such as IL-6, L-1β and TNF-α, iMC differentiate into MDSC. Furthermore, cancer cells secrete factors including PGE2 and CXCL12 that help in the recruitment of MDSC to the TME. Finally, the activation of STAT3 pathway results in the expression of several factors such as ARG-1, IDO, TGF-β, ROS, etc. These are involved in mediating the tumor promoting function of MDSC. Abbreviations: *ARG-1* arginase-1; *CXCL12* chemokine C-X-C motif ligand 12; *G-CSF* granulocyte-colony stimulating factor; *GM-CSF* granulocyte macrophage-CSF; *HSC* hematopoietic stem cell; *IDO* indoleamine 2,3 deoxygenase; *IL* interleukin; *iMC* immature myeloid cell; *M-CSF* macrophage-CSF; *MDSC* myeloid-derived suppressor cell; *PC* progenitor cell; *PGE2* prostaglandin E2; *ROS* reactive oxygen species; *STAT3* signal transducer and activator of transcription 3; *TGF-β* transforming growth factor-β; *TME* tumor microenvironment; *TNF-α* tumor necrosis factor-α

they process the exogenously acquired tumor antigen into peptides. These are presented in major histocompatibility complex (MHC) class II molecules to the T-cell receptor (TCR) of CD4$^+$ T cells. Moreover, certain APC subsets such as CD8α$^+$ DC are able to shuttle peptides from exogenously acquired tumor antigens to the MHC I pathway, as such enabling cross-priming of CD8$^+$ T cells [2, 3]. Activation of CD4$^+$ T helper 1 cells (T$_H$1) and CD8$^+$ cytotoxic T lymphocytes (CTL) by APC occurs when the tumor antigen is acquired in the context of "danger" as this provides signals to the APC to become fully stimulatory cells [4]. Once activated, T$_H$1 and CTL egress from the lymph node in search of tumor cells that present their cognate antigen. At the tumor site, T$_H$1 and CTL act in concert to kill cancer cells [5]. Tumor cells however adapt at escaping T-cell killing [6] and while some tumor cells are successfully eliminated, other more harmful variants can escape the initial immune attack. During their development, cancers become highly infiltrated with different subsets of immune cells of lymphoid and myeloid origin that exert suppressive activities. These subsets are referred to as regulatory immune cells and include regulatory T cells (Treg), tumor-associated DC (TADC), tumor-associated macrophages (TAM), and myeloid-derived suppressor cells (MDSC). Together with tumor cells, regulatory immune cells actively quench T-cell activity and enable cancer cells to grow undisturbed [7].

The main goal of cancer immunologists is to empower CTL to reject cancer cells. Much attention has been devoted to the stimulation of tumor-specific CTL. However, it has become increasingly clear that a cancer-curing immunotherapy also has to interfere with pathways that affect their function at the TME. Defining which inhibitory pathway is a "universal obstacle" and as such an ideal target has been a challenging endeavor as the infiltration of suppressive immune cells and the mechanisms they exert can vary considerably between cancer types. In this regard, MDSC have come to the forefront as a target population, because they are prevalent in most cancer patients and because they exploit a plethora of mechanisms to directly or indirectly abrogate antitumor immunity [8, 9]. However, the heterogeneity of MDSC and the diversity of inhibitory mechanisms they employ have faced us with the challenge of finding a "one fits all" strategy to deplete and/or functionally modulate them. Fortunately, the behavior of a cell in large is dictated by transcriptional programs. In the case of MDSC, it has been suggested that the transcription factor signal transducer and activator of transcription 3 (STAT3) is a main regulator [9–15]. This is further highlighted by the observation that STAT3 expressed by MDSC is implied in their accumulation, differentiation, and functionality (Fig. 6.1).

6.2 Signal Transducer and Activator of Transcription 3: The Basics

The STAT family is comprised of seven members that are encoded by distinct genes. Similar to its other family members, STAT3 is present in non-stimulated cells in an inactive cytoplasmic form. Activation of STAT3 can be triggered through a multitude of factors amongst which are interleukin-6 (IL-6)-like cytokines [16], colony-stimulating factors (CSF), leptin [17], interferon (IFN) family members, IL-2 family members, and growth factors such as epidermal growth factor [18]. Depending on the trigger, STAT3 activation occurs through its phosphorylation on tyrosine 705 or serine 727. Phosphorylation on tyrosine 705 can be regulated by different tyrosine kinases and by members of the Janus-activated kinases (JAK) [19], whereas phosphorylation of serine 727 can be regulated by protein kinase C, mitogen-activated protein kinases, and cyclin-dependent kinase 5 [15]. Phosphorylation of STAT3 results in its dimerization, which enables STAT3 to act as a transcriptional activator of various target genes. Also acetylation of lysine 685 has been described as a mode of STAT3 activation [20] and a way to enhance the stability of STAT3 dimers [15]. All transcriptional activity requires tight control, which in the case of STAT3 is performed by various negative regulators such as protein inhibitor of activated STAT (PIAS) proteins [21], suppressors of cytokine signaling (SOCS) proteins [22], and protein tyrosine phosphatases [23, 24]. These families of STAT3 regulating proteins interfere with STAT3 binding to DNA, hamper tyrosine kinases and remove phosphates from activated STAT3, respectively. In addition, STAT3 levels can be regulated through ubiquitination-dependent proteosomal degradation [25].

A large body of evidence has shown that STAT3 is constitutively activated in many mouse tumor models [13, 26–28] and more importantly in human cancers including breast, liver, lung, pancreas, prostate, skin, and hematological cancers [29–35]. This is explained by the fact that many of the triggers that activate STAT3 are abundantly present in the TME. Moreover, a number of genes induced by STAT3 provide a positive feedback and as such keep the STAT3 pathway continuously activated. Importantly, STAT3 activation occurs in both cancer cells and the many immune cells that infiltrate tumors, including MDSC [9–15]. It has been described that STAT3 is one of the factors that allows crosstalk between the different cell types that are part of the TME and therefore represents an attractive target for modulation. Although activated STAT3 is not only linked to the life cycle of MDSC, we will limit the remaining of this book chapter to the role of STAT3 on MDSC accumulation and function (Fig. 6.1), since in contrast to other tumor-infiltrating immune cells such as TADC, TAM, and T cells, MDSC are abundantly present in most mouse tumor models and human cancers [8, 9]. Moreover, it is becoming increasingly clear that many cancer therapies, amongst which anticancer vaccination, are more effective when MDSC have been depleted [36–42].

6.3   Signal Transducer and Activator of Transcription 3 Play a Role in the Accumulation and Differentiation of Myeloid-Derived Suppressor Cells

Although MDSC were described in 1970 as natural suppressor cells [43], it took until 2007 for the term MDSC to get established. Generally, the name MDSC is used to categorize a heterogeneous mix of immature myeloid cells, which can be found in various pathological conditions including cancer [44]. In healthy individuals' immature myeloid cells, which differentiate into mature macrophages, DCs and granulocytes are constantly generated in the bone marrow. In cancer-bearing subjects, the differentiation of immature myeloid cells is disturbed through the presence of many tumor-derived factors that favor immature myeloid cell accumulation and differentiation to MDSC both at the tumor site and secondary lymphoid organs [45]. Cancer-derived factors that drive the generation of MDSC in the bone marrow include granulocyte-colony stimulating factor (G-CSF) and granulocyte macrophage-CSF (GM-CSF), various interleukins like IL-6 and IL-1β, prostaglandin E2 (PGE2), tumor necrosis factor-α (TNF-α), and vascular endothelial growth factor (VEGF) [46]. Many of these activate the STAT3 pathway, so it is no surprise that STAT3 signaling has been implicated in the stimulation of myeloid cell differentiation into MDSC. STAT3 has been shown to interact with CCAAT-enhancer-binding protein β (C/EBPβ), a transcription factor that has a key role in myeloid development [47]. Importantly, bone-marrow cells deficient for C/EBPβ lose the ability to differentiate into functional MDSC [47]. Furthermore, a correlation between C/EBPβ and accumulation of CD11b$^+$ Gr-1$^+$ cells in response to G-CSF was reported [48, 49]. This observation and the finding that STAT3 deficiency makes myeloid progenitors refractory to growth stimulation by G-CSF [50], suggest that STAT3 and C/EBPβ are inextricably linked to MDSC generation. This is further supported by the observation that STAT3 prolongs the binding of C/EBPβ on the myc promoter [47]. Besides myc, other cell survival and cell cycle regulating proteins including Bcl-xL, survivin, and cyclin D1 are upregulated by STAT3 [8, 15], as well as multiple other proteins critical for MDSC differentiation such as S100A proteins [51] and protein kinase C βII (PKCβII) [52]. The latter two inhibit DC differentiation from myeloid progenitor cells and thereby promote MDSC accumulation. The studies described above clearly point towards a role for STAT3 in the expansion and differentiation of hematopoietic stem cells and myeloid progenitor cells into MDSC.

6.4 Activated Signal Transducer and Activator of Transcription 3 can be Found in Monocytic and Polymorphonuclear Myeloid-Derived Suppressor Cells

MDSC can be found at elevated levels in the periphery and are recruited to the TME through the secretion of chemokines such as chemokine C-C motif ligand 2 (CCL2), chemokine C-X-C motif ligand 5 (CXCL5), and CXCL12 [46], as well as other factors including IL-6, IL-1β, G-CSF, and VEGF [53]. In mice, MDSC are defined as CD11b and Gr-1 expressing cells [54]. Antibodies recognizing the granulocyte-specific marker Gr-1 target an epitope that is shared amongst the antigens Ly6C and Ly6G, two markers that have been used to divide MDSC in monocytic (MO)-MDSC or Ly6C^{high}Ly6G^{low} cells, and polymorphonuclear (PMN)-MDSC or Ly6C^{low}Ly6G^{high} cells [55]. Corresponding populations have been described in cancer patients. In general, human MDSC are characterized by the expression of CD33, CD11b and the absence of significant levels of other lineage markers and HLA-DR [56]. Human MO-MDSC are further characterized as CD14$^+$ but CD15$^-$ cells, while human PMN-MDSC are defined as CD14$^-$ CD15$^+$ [9]. Although these phenotypes have generally been accepted to define MDSC, several other surface markers have been put forward to distinguish MDSC subsets based on their function, amongst others CD40, CD49 (VLA4), CD80 (B7.1), CD115 (M-CSFR), and CD124 (IL4Rα) [10, 57–62]. Although these markers are undoubtedly expressed on MDSC, it is generally accepted that they do not define specific MDSC subsets [63], and that at least the expression of some markers such as CD80 varies considerably depending on the cancer type and MDSC location [64]. Because of this phenotypic heterogeneity, it has frequently been suggested that the suppressive activity of MDSC is the ultimate defining characteristic [65]. The latter is in part dictated by the activation of STAT3 in MDSC. Of note, expression of phosphorylated STAT3 has been described in both subsets.

6.5 Signal Transducer and Activator of Transcription 3 and its Role in the Tumor Promoting Activity of Myeloid-Derived Suppressor Cells

Several mechanisms are employed by MDSC to promote tumor growth including suppression of antitumor responses, stimulation of angiogenesis as well as tumor cell metastasis. These activities have been linked to activation of STAT3 in the MDSC.

Immunosuppression is an important biological characteristic of MDSC. To that end, MDSC deplete nutrients required by T cells, generate oxidative stress, activate and expand Treg, and finally inhibit T-cell trafficking [8]. Several mechanisms that are at the basis of these MDSC activities have been linked to phosphorylation of STAT3. For instance, expression of arginase-1 (ARG-1) is under the control of STAT3 and results in consumption of L-arginine and L-cysteine [9, 66–70]. Depletion of these amino acids results in downregulation of the CD3ζ-chain in the TCR complex and growth arrest of antigen-activated T cells [71, 72]. Moreover, Serafini et al. [73] linked the expression of ARG-1 to expansion of Treg by MDSC in a B-cell lymphoma model. In this model, transforming growth factor-β (TGF-β) produced by the MDSC had no effect on Treg. Nonetheless, TGF-β has been linked to T-cell suppression [74] and Treg expansion [75, 76]. Importantly, a link between TGF-β and STAT3 was proposed based on the presence of two STAT3 binding sites in the TGF-β promoter [77]. Moreover, it was shown that TGF-β production was reduced after myeloid-specific STAT3 knock down [75]. This reduction in TGF-β was correlated to a reduction in Treg numbers. Another enzyme that is under the control of STAT3 is indoleamine 2,3 deoxygenase (IDO) [78]. This enzyme depletes tryptophan, thereby generating the toxic metabolite kynurenine. The mode of action of IDO is similar to that of ARG-1, suppression of TCR-mediated effector T-cell activation, growth arrest and induction of effector T-cell apoptosis, and expansion of Treg [79, 80]. Besides depletion of nutrients required for T cells and Treg expansion, STAT3 has also been linked to the generation of oxidative stress, another mechanism that dampens antitumor immunity [81]. Elevated production of reactive oxygen species (ROS) has been linked in a variety of mouse cancer models and human cancers to the enhanced expression of NAPDH oxidase, which in turn is controlled by STAT3 [15, 81]. It was postulated that S100A8/A9 hetrodimers assist in the formation of the NADPH oxidase complex [50]. Moreover, ARG-1 can also contribute to ROS production [82, 83]. Importantly, ROS play a role in the suppression of antigen-specific T cells [84–86] and have been shown to induce T-cell apoptosis [87], much in the same way as ARG-1. These studies show a central role for STAT3 in the active quenching of antitumor immunity by MDSC.

Immunosuppression is not the only way by which MDSC support tumor growth. These cells also promote tumor progression by enhancing blood vessel development, tumor cell invasion, and metastasis. Angiogenesis has been linked to enhanced production of VEGF and basic fibroblast growth factor (bFGF) by MDSC. These angiogenic factors are under the control of STAT3 [88]. Moreover, STAT3-driven proteases such as metaloproteinase 9 (MMP9), and TGF-β have also been linked to angiogenesis [89]. In this regard, MMP9 was shown to enhance the bioavailability of VEGF and as such support vascular stability [90]. In addition to a role in vasculogenesis, MMPs also play a role in promoting tumor cell metastasis. Furthermore, MDSC expressing active STAT3 have been implicated in the formation of pre-metastatic niches [91, 92]. These cells condition organs by creating an immunosuppressive environment that allows growth of metastatic tumor cells [93–95]. Herein, STAT3-regulated factors such as bFGF, IL, MMP9, and S100A proteins play a role [93, 96]. It was recently shown that CD8$^+$ T cells are able to induce MDSC apoptosis at distant sites and as such might inhibit MDSC accumulation in pre-metastatic niches. However, activation of STAT3 compromises the ability of effector T cells to kill MDSC [91, 92]. This was linked to a lower granzyme B expression by CD8$^+$ T cells and resistance of MDSC to T-cell killing. Importantly, Zhang et al. [91] showed a positive correlation between STAT3 activation and myeloid cell accumulation, increased IL-10, IL-6, and VEGF, while they observed an inverse correlation between STAT3 activation and CD8$^+$ T cell numbers as well as the expression of granzyme B by T cells in melanoma draining lymph nodes. These data from patients highlight the relevance of the mouse study and further point towards STAT3 as a master regulator of the MDSC's tumor promoting activity.

6.6 Targeting Signal Transducer and Activator of Transcription 3 as a Strategy to Manipulate Myeloid-Derived Suppressor Cells

As mentioned previously, MDSC have come to the forefront as a target in cancer (immuno) therapy because of several reasons. Firstly, MDSC are abundantly present in most cancer patients, irrespective of the cancer type [9]. Secondly, the presence of MDSC correlates with cancer stage and metastatic disease [97]. Thirdly, MDSC accelerate tumor progression by inhibiting antitumor immune responses, stimulating angiogenesis and tumor cell metastasis [8, 9]. Throughout this book chapter, we showed that STAT3 is implicated in the accumulation, differentiation, and function of MDSC. Consequently, several research teams have evaluated STAT3 targeting drugs as a means to interfere with these processes and as such put a brake on tumor progression [39, 98]. The list of drugs includes curcumin derivatives and other JAK2/STAT3 inhibitors including AZD1480 [14, 99–102], Icariin flavone and its derivative 3,5,7-trihydroxy-4'-emthoxy-8-(3-hydroxy-3-methylbutyl)-flavone [103], tyrosine kinase inhibitors including sunitinib [104–107], VEGF inhibiting molecules such as VEGF-trap (a VEGF receptor fused to the Fc part of human IgG1) [108, 109] and anti-VEGF antibodies (bevacizumab) [110, 111], monoclonal antibodies specific for IL-6 [112], and molecules such as bardoxolone methyl (CDDO-Me) [37, 113–115].

Curcurmin and its derivatives are naturally occurring phenols that are used for their anti-oxidant and anti-inflammatory activities. Furthermore, these have been used to selectively inhibit the JAK2/STAT3 pathway [99–102]. Administration of cucurbitacin B (CuB) to lung cancer patients was shown to decrease the numbers of bona fide MDSC (Lin$^-$ HLA-DR$^-$ CD33$^+$), while it increased the numbers of mature Lin$^-$ HLA-DR$^+$ CD33$^+$ myeloid cells in peripheral blood. Moreover, it was shown in vitro that CuB induced DC differentiation and increased the sensitivity of tumor cells to antigen (p53)-specific T cells [100]. Other JAK2/STAT3 inhibitors have been tested including AZD1480 [14]. In a study on mouse MDSC, it was shown that AZD1480 has a direct effect on the levels of MDSC. However, it did not abrogate the ability of MDSCs to suppress T cells. In contrast, when evaluated on a per cell basis, it was shown that the suppressive activity of the MDSC was higher after treatment with AZD1480. Similar to JAK2/STAT3 inhibitors, flavanoids such as Icariin and its derivative were reported to downregulate MDSC numbers [103]. These natural compounds were shown to inhibit STAT3 signaling and expression of S100A8 and S100A9, resulting in differentiation of immature myeloid cells to mature cells. Sunitinib is a small-molecule multikinase inhibitor that targets amongst others the VEGF receptor, platelet-derived growth factor receptor and c-kit, and as such hampers the phosphorylation of STAT3. Ko et al. [104] showed that sunitinib efficiently eliminates peripheral MDSC, whereas it did not reduce MDSC in tumors. This was linked to high levels of GM-CSF in the tumor and STAT5 signaling in MDSC. Nonetheless, other studies show that MDSC depletion by sunitinib is irrespective of the location [106]. Importantly, treatment of metastatic renal cell cancer patients with sunitinib reduced the level of MDSC in peripheral blood by half and was associated with improved T_H1 function (reduced IL-4 and higher interferon-γ [IFN-γ]) and lower Treg numbers [105, 107]. Although sunitinib, which affects downstream VEGF receptor signaling and as such STAT3 activation, was shown to modulate MDSC levels, other strategies that impact on VEFG receptor signaling, such as VEGF-trap [108, 109] and anti-VEGF antibodies demonstrated no effect on MDSC levels in peripheral blood of cancer patients [111]. This is an unexpected finding, since the link between VEGF and MDSC accumulation is longstanding and as it was shown that anti-VEGF antibodies successfully reduce MDSC numbers in mice [110]. Besides antibodies to capture VEGF and as such inhibit STAT3 activation upon VEGF receptor interaction, researchers have developed monoclonal antagonsitic antibodies specific for the IL-6 receptor, as its triggering is directly linked to STAT3 activation and MDSC. These anti-IL-6 receptor antibodies neutralize tumor-derived IL-6 and suppress expansion of cancer-associated MDSC [112]. Finally, a molecule that was shown to inhibit STAT3 activation in MDSC, at least when used at high concentrations (1-5 μM), is CDDO-Me. Treatment with this synthetic triterpentoid (a methyl ester of 2-cyano-3,12-dioxooleana-1,9 (11)-dien-28-oic acid) resulted in reduced production of ROS, improved T-cell function, and more importantly reduced tumor growth [37, 113–115].

The studies above show the potential of targeting STAT3 in MDSC as an anticancer strategy. At the same time, these studies show that although the aforementioned drugs act on STAT3 activation, their mode of action can differ from MDSC depletion, maturation to functional modulation. Moreover, treatment of cancer-bearing subjects with only these drugs was shown to be insufficient to provide a cure. As MDSC represent a confounding factor for antitumor immunity and as it was shown that MDSC depletion improves the outcome of cancer vaccines [36–42], it is not surprising that drugs such as CDDMO-Me [37] and sunitinib [41, 42, 116] have been evaluated in combination with cancer vaccination. In these studies, the combination therapy showed improved curative potential when compared to either component alone. However, instead of combining therapies it would be more elegant if one drug could lead to activation of tumor-specific CTLs while inhibiting MDSC. Importantly, various studies have shown that MDSC can be reverted into stimulatory APC under the influence of cytokines such as IL-12 [117, 118] or Toll-like receptor (TLR) ligands such as CpG oligonucleotides [119, 120]. The latter has offered an opportunity to design a drug consisting of CpG oligonucleotides conjugated to STAT3-specific small interfering RNA (referred to as CpG-siSTAT3 conjugates) [121–123]. It has been shown in mouse cancer models and using STAT3[+] PMN-MDSC of prostate cancer patients that CpG-siSTAT3 conjugates mediate selective delivery of silencing siSTAT3 to TLR9[+] myeloid cells, resulting in disruption of the STAT3-regulated suppressive signaling network and stimulation of antitumor immunity. These findings indicate that this gene- and cell type-specific inhibitory oligonucleotides represent novel therapeutic approaches to mitigate immunosuppression in cancer patients.

6.7   Conclusions

There is ample evidence on the association of MDSC to poor prognosis both in primary and metastatic tumors regardless of the heterogeneity of MDSC, particularly in cancer patients. This observation provides a rationale for therapeutic approaches that target MDSC. But which property of MDSC should we target, their ability to suppress antitumor immunity, to stimulate vascularization or to stimulate tumor cell metastasis? Or should we try to deplete these cells all together? Well ...
As STAT3 is a major regulator of MDSC accumulation, differentiation, and function, it offers us an opportunity to develop a strategy that allows us to reprogram MDSC into mature myeloid cells that counteract tumor growth amongst others through the stimulation of antitumor immunity. From the various studies on STAT3 targeted drugs, their impact on MDSC and tumor growth, it has become clear that we should develop a drug that targets STAT3 in MDSC specifically. This is underscored by the data provided by the CpG-siSTAT3 conjugate studies, which show that targeted delivery of STAT3 inhibiting molecules is the way forward [120–123]. Most likely, a multidisciplinary approach will be required to design MDSC-targeted STAT3 inhibitors with a commercial profile. Such an approach could encompass the use of myeloid cell-targeted lentiviral vectors [124, 125] or nanoparticles [126], which could then deliver silencing RNA for STAT3 [13, 121–123] or genes encoding negative regulators of STAT3 [21–24].

References

1. Coulie PG, Van den Eynde BJ, van der Bruggen P, Boon T (2014) Tumour antigens recognized by T lymphocytes: at the core of cancer immunotherapy. Nat Rev Cancer 14 (2):135–146. doi:10.1038/nrc3670 nrc3670 [pii]
2. Brossart P, Bevan MJ (1997) Presentation of exogenous protein antigens on major histocompatibility complex class I molecules by dendritic cells: pathway of presentation and regulation by cytokines. Blood 90(4):1594–1599
3. Hildner K, Edelson BT, Purtha WE, Diamond M, Matsushita H, Kohyama M, Calderon B, Schraml BU, Unanue ER, Diamond MS, Schreiber RD, Murphy TL, Murphy KM (2008) Batf3 deficiency reveals a critical role for CD8alpha + dendritic cells in cytotoxic T cell immunity. Science 322(5904):1097–1100. doi:10.1126/science.1164206 322/5904/1097 [pii]
4. Pradeu T, Cooper EL (2012) The danger theory: 20 years later. Front Immunol 3:287. doi:10.3389/fimmu.2012.00287
5. Bonehill A, Heirman C, Thielemans K (2005) Genetic approaches for the induction of a CD4 + T cell response in cancer immunotherapy. J Gene Med 7(6):686–695. doi:10.1002/jgm.713
6. Hanahan D, Weinberg RA (2011) Hallmarks of cancer: the next generation. Cell 144 (5):646–674. doi:10.1016/j.cell.2011.02.013 S0092-8674(11)00127-9 [pii]
7. Dunn GP, Bruce AT, Ikeda H, Old LJ, Schreiber RD (2002) Cancer immunoediting: from immunosurveillance to tumor escape. Nat Immunol 3(11):991–998. doi:10.1038/ni1102-991 ni1102-991 [pii]
8. Gabrilovich DI, Ostrand-Rosenberg S, Bronte V (2012) Coordinated regulation of myeloid cells by tumours. Nat Rev Immunol 12(4):253–268. doi:10.1038/nri3175

9. Jiang J, Guo W, Liang X (2014) Phenotypes, accumulation, and functions of myeloid-derived suppressor cells and associated treatment strategies in cancer patients. Hum Immunol 75(11):1128–1137. doi:10.1016/j.humimm.2014.09.025 S0198-8859(14) 00465-0 [pii]

10. Poschke I, Mougiakakos D, Hansson J, Masucci GV, Kiessling R (2010) Immature immunosuppressive CD14 + HLA-DR-/low cells in melanoma patients are Stat3hi and overexpress CD80, CD83, and DC-sign. Cancer Res 70(11):4335–4345. doi:10.1158/0008-5472.CAN-09-3767 0008-5472.CAN-09-3767 [pii]

11. Nefedova Y, Huang M, Kusmartsev S, Bhattacharya R, Cheng P, Salup R, Jove R, Gabrilovich D (2004) Hyperactivation of STAT3 is involved in abnormal differentiation of dendritic cells in cancer. J Immunol 172(1):464–474

12. Cheng P, Corzo CA, Luetteke N, Yu B, Nagaraj S, Bui MM, Ortiz M, Nacken W, Sorg C, Vogl T, Roth J, Gabrilovich DI (2008) Inhibition of dendritic cell differentiation and accumulation of myeloid-derived suppressor cells in cancer is regulated by S100A9 protein. J Exp Med 205(10):2235–2249. doi:10.1084/jem.20080132 jem.20080132 [pii]

13. Emeagi PU, Maenhout S, Dang N, Heirman C, Thielemans K, Breckpot K (2013) Downregulation of Stat3 in melanoma: reprogramming the immune microenvironment as an anticancer therapeutic strategy. Gene Ther 20(11):1085–1092. doi:10.1038/gt.2013.35 gt201335 [pii]

14. Maenhout SK, Du Four S, Corthals J, Neyns B, Thielemans K, Aerts JL (2014) AZD1480 delays tumor growth in a melanoma model while enhancing the suppressive activity of myeloid-derived suppressor cells. Oncotarget 5(16):6801–6815. doi:10.18632/oncotarget. 2254 2254 [pii]

15. Rebe C, Vegran F, Berger H, Ghiringhelli F (2013) STAT3 activation: A key factor in tumor immunoescape. JAKSTAT 2(1):e23010. doi:10.4161/jkst.23010 2012JAKS0054R [pii]

16. Kishimoto T, Akira S, Narazaki M, Taga T (1995) Interleukin-6 family of cytokines and gp130. Blood 86(4):1243–1254

17. Akira S (1999) Functional roles of STAT family proteins: lessons from knockout mice. Stem Cells 17(3):138–146. doi:10.1002/stem.170138

18. Zhong Z, Wen Z, Darnell JE Jr (1994) Stat3: a STAT family member activated by tyrosine phosphorylation in response to epidermal growth factor and interleukin-6. Science 264 (5155):95–98

19. Hirano T, Ishihara K, Hibi M (2000) Roles of STAT3 in mediating the cell growth, differentiation and survival signals relayed through the IL-6 family of cytokine receptors. Oncogene 19(21):2548–2556. doi:10.1038/sj.onc.1203551

20. Yuan ZL, Guan YJ, Chatterjee D, Chin YE (2005) Stat3 dimerization regulated by reversible acetylation of a single lysine residue. Science 307(5707):269–273. doi:10.1126/science. 1105166 307/5707/269 [pii]

21. Shuai K, Liu B (2005) Regulation of gene-activation pathways by PIAS proteins in the immune system. Nat Rev Immunol 5(8):593–605. doi:10.1038/nri1667 nri1667 [pii]

22. Alexander WS, Hilton DJ (2004) The role of suppressors of cytokine signaling (SOCS) proteins in regulation of the immune response. Annu Rev Immunol 22:503–529. doi:10. 1146/annurev.immunol.22.091003.090312

23. Irie-Sasaki J, Sasaki T, Matsumoto W, Opavsky A, Cheng M, Welstead G, Griffiths E, Krawczyk C, Richardson CD, Aitken K, Iscove N, Koretzky G, Johnson P, Liu P, Rothstein DM, Penninger JM (2001) CD45 is a JAK phosphatase and negatively regulates cytokine receptor signalling. Nature 409(6818):349–354. doi:10.1038/35053086

24. Schaper F, Gendo C, Eck M, Schmitz J, Grimm C, Anhuf D, Kerr IM, Heinrich PC (1998) Activation of the protein tyrosine phosphatase SHP2 via the interleukin-6 signal transducing receptor protein gp130 requires tyrosine kinase Jak1 and limits acute-phase protein expression. Biochem J 335(Pt 3):557–565

25. Daino H, Matsumura I, Takada K, Odajima J, Tanaka H, Ueda S, Shibayama H, Ikeda H, Hibi M, Machii T, Hirano T, Kanakura Y (2000) Induction of apoptosis by extracellular

ubiquitin in human hematopoietic cells: possible involvement of STAT3 degradation by proteasome pathway in interleukin 6-dependent hematopoietic cells. Blood 95(8):2577–2585

26. Yang CH, Fan M, Slominski AT, Yue J (2010) Pfeffer LM (2010) The role of constitutively activated STAT3 in B16 melanoma cells. Int J Infereron Cytokine Mediator Res 2:1–7. doi:10.2147/IJICMR.S6657

27. Liu Y, Deng J, Wang L, Lee H, Armstrong B, Scuto A, Kowolik C, Weiss LM, Forman S, Yu H (2012) S1PR1 is an effective target to block STAT3 signaling in activated B cell-like diffuse large B-cell lymphoma. Blood 120(7):1458–1465. doi:10.1182/blood-2011-12-399030 blood-2011-12-399030 [pii]

28. Tkach M, Coria L, Rosemblit C, Rivas MA, Proietti CJ, Diaz Flaque MC, Beguelin W, Frahm I, Charreau EH, Cassataro J, Elizalde PV, Schillaci R (2012) Targeting Stat3 induces senescence in tumor cells and elicits prophylactic and therapeutic immune responses against breast cancer growth mediated by NK cells and CD4 + T cells. J Immunol 189(3):1162–1172. doi:10.4049/jimmunol.1102538 jimmunol.1102538 [pii]

29. Buettner R, Mora LB, Jove R (2002) Activated STAT signaling in human tumors provides novel molecular targets for therapeutic intervention. Clin Cancer Res 8(4):945–954

30. Bowman T, Garcia R, Turkson J, Jove R (2000) STATs in oncogenesis. Oncogene 19 (21):2474–2488. doi:10.1038/sj.onc.1203527

31. Krasilnikov M, Ivanov VN, Dong J, Ronai Z (2003) ERK and PI3 K negatively regulate STAT-transcriptional activities in human melanoma cells: implications towards sensitization to apoptosis. Oncogene 22(26):4092–4101. doi:10.1038/sj.onc.1206598 1206598 [pii]

32. Darnell JE (2005) Validating Stat3 in cancer therapy. Nat Med 11(6):595–596. doi:10.1038/nm0605-595 nm0605-595 [pii]

33. Hung MH, Tai WT, Shiau CW, Chen KF (2014) Downregulation of signal transducer and activator of transcription 3 by sorafenib: a novel mechanism for hepatocellular carcinoma therapy. World J Gastroenterol 20(41):15269–15274. doi:10.3748/wjg.v20.i41.15269

34. Scuto A, Kujawski M, Kowolik C, Krymskaya L, Wang L, Weiss LM, Digiusto D, Yu H, Forman S, Jove R (2011) STAT3 inhibition is a therapeutic strategy for ABC-like diffuse large B-cell lymphoma. Cancer Res 71(9):3182–3188. doi:10.1158/0008-5472.CAN-10-2380 0008-5472.CAN-10-2380 [pii]

35. Catlett-Falcone R, Landowski TH, Oshiro MM, Turkson J, Levitzki A, Savino R, Ciliberto G, Moscinski L, Fernandez-Luna JL, Nunez G, Dalton WS, Jove R (1999) Constitutive activation of Stat3 signaling confers resistance to apoptosis in human U266 myeloma cells. Immunity 10(1):105–115. doi:10.1016/S1074-7613(00)80011-4 [pii]

36. Kusmartsev S, Cheng F, Yu B, Nefedova Y, Sotomayor E, Lush R, Gabrilovich D (2003) All-trans-retinoic acid eliminates immature myeloid cells from tumor-bearing mice and improves the effect of vaccination. Cancer Res 63(15):4441–4449

37. Nagaraj S, Youn JI, Weber H, Iclozan C, Lu L, Cotter MJ, Meyer C, Becerra CR, Fishman M, Antonia S, Sporn MB, Liby KT, Rawal B, Lee JH, Gabrilovich DI (2010) Anti-inflammatory triterpenoid blocks immune suppressive function of MDSCs and improves immune response in cancer. Clin Cancer Res 16(6):1812–1823. doi:10.1158/1078-0432.CCR-09-3272 1078-0432.CCR-09-3272 [pii]

38. Iclozan C, Antonia S, Chiappori A, Chen DT, Gabrilovich D (2013) Therapeutic regulation of myeloid-derived suppressor cells and immune response to cancer vaccine in patients with extensive stage small cell lung cancer. Cancer Immunol Immunother 62(5):909–918. doi:10.1007/s00262-013-1396-8

39. Wesolowski R, Markowitz J, Carson WE 3rd (2013) Myeloid derived suppressor cells - a new therapeutic target in the treatment of cancer. Journal for immunotherapy of cancer 1:10. doi:10.1186/2051-1426-1-10

40. Poschke I, Mao Y, Adamson L, Salazar-Onfray F, Masucci G, Kiessling R (2012) Myeloid-derived suppressor cells impair the quality of dendritic cell vaccines. Cancer Immunol Immunother 61(6):827–838. doi:10.1007/s00262-011-1143-y

41. Bose A, Taylor JL, Alber S, Watkins SC, Garcia JA, Rini BI, Ko JS, Cohen PA, Finke JH, Storkus WJ (2011) Sunitinib facilitates the activation and recruitment of therapeutic

anti-tumor immunity in concert with specific vaccination. Int J Cancer 129(9):2158–2170. doi:10.1002/ijc.25863

42. Draghiciu O, Nijman HW, Hoogeboom BN, Meijerhof T, Daemen T (2015) Sunitinib depletes myeloid-derived suppressor cells and synergizes with a cancer vaccine to enhance antigen-specific immune responses and tumor eradication. Oncoimmunology 4(3):e989764. doi:10.4161/2162402X.2014.989764 989764 [pii]

43. Strober S (1984) Natural suppressor (NS) cells, neonatal tolerance, and total lymphoid irradiation: exploring obscure relationships. Annu Rev Immunol 2:219–237. doi:10.1146/annurev.iy.02.040184.001251

44. Gabrilovich DI, Bronte V, Chen SH, Colombo MP, Ochoa A, Ostrand-Rosenberg S, Schreiber H (2007) The terminology issue for myeloid-derived suppressor cells. Cancer Res 67(1):425, author reply 426. doi:10.1158/0008-5472.CAN-06-3037 67/1/425 [pii]

45. Bronte V, Serafini P, Apolloni E, Zanovello P (2001) Tumor-induced immune dysfunctions caused by myeloid suppressor cells. J Immunother 24(6):431–446

46. Sawanobori Y, Ueha S, Kurachi M, Shimaoka T, Talmadge JE, Abe J, Shono Y, Kitabatake M, Kakimi K, Mukaida N, Matsushima K (2008) Chemokine-mediated rapid turnover of myeloid-derived suppressor cells in tumor-bearing mice. Blood 111 (12):5457–5466. doi:10.1182/blood-2008-01-136895 blood-2008-01-136895 [pii]

47. Zhang H, Nguyen-Jackson H, Panopoulos AD, Li HS, Murray PJ, Watowich SS (2010) STAT3 controls myeloid progenitor growth during emergency granulopoiesis. Blood 116 (14):2462–2471. doi:10.1182/blood-2009-12-259630 blood-2009-12-259630 [pii]

48. Marigo I, Bosio E, Solito S, Mesa C, Fernandez A, Dolcetti L, Ugel S, Sonda N, Bicciato S, Falisi E, Calabrese F, Basso G, Zanovello P, Cozzi E, Mandruzzato S, Bronte V (2010) Tumor-induced tolerance and immune suppression depend on the C/EBPbeta transcription factor. Immunity 32(6):790–802. doi:10.1016/j.immuni.2010.05.010 S1074-7613(10)00202-5 [pii]

49. Hirai H, Zhang P, Dayaram T, Hetherington CJ, Mizuno S, Imanishi J, Akashi K, Tenen DG (2006) C/EBPbeta is required for 'emergency' granulopoiesis. Nat Immunol 7(7):732–739. doi:10.1038/ni1354 ni1354 [pii]

50. Trikha P, Carson WE (1846) 3rd (2014) Signaling pathways involved in MDSC regulation. Biochim Biophys Acta 1:55–65. doi:10.1016/j.bbcan.2014.04.003 S0304-419X(14)00039-0 [pii]

51. Foell D, Wittkowski H, Vogl T, Roth J (2007) S100 proteins expressed in phagocytes: a novel group of damage-associated molecular pattern molecules. J Leukoc Biol 81(1):28–37. doi:10.1189/jlb.0306170 jlb.0306170 [pii]

52. Farren MR, Carlson LM, Lee KP (2010) Tumor-mediated inhibition of dendritic cell differentiation is mediated by down regulation of protein kinase C beta II expression. Immunol Res 46(1–3):165–176. doi:10.1007/s12026-009-8118-5

53. Poschke I, Kiessling R (2012) On the armament and appearances of human myeloid-derived suppressor cells. Clin Immunol 144(3):250–268. doi:10.1016/j.clim.2012.06.003 S1521-6616(12)00157-X [pii]

54. Bronte V, Wang M, Overwijk WW, Surman DR, Pericle F, Rosenberg SA, Restifo NP (1998) Apoptotic death of CD8 + T lymphocytes after immunization: induction of a suppressive population of Mac-1 +/Gr-1 + cells. J Immunol 161(10):5313–5320

55. Movahedi K, Guilliams M, Van den Bossche J, Van den Bergh R, Gysemans C, Beschin A, De Baetselier P, Van Ginderachter JA (2008) Identification of discrete tumor-induced myeloid-derived suppressor cell subpopulations with distinct T cell-suppressive activity. Blood 111(8):4233–4244. doi:10.1182/blood-2007-07-099226 blood-2007-07-099226 [pii]

56. Almand B, Clark JI, Nikitina E, van Beynen J, English NR, Knight SC, Carbone DP, Gabrilovich DI (2001) Increased production of immature myeloid cells in cancer patients: a mechanism of immunosuppression in cancer. J Immunol 166(1):678–689

57. Huang B, Pan PY, Li Q, Sato AI, Levy DE, Bromberg J, Divino CM, Chen SH (2006) Gr-1 + CD115 + immature myeloid suppressor cells mediate the development of

tumor-induced T regulatory cells and T-cell anergy in tumor-bearing host. Cancer Res 66 (2):1123–1131. doi:10.1158/0008-5472.CAN-05-1299 66/2/1123 [pii]

58. Pan PY, Ma G, Weber KJ, Ozao-Choy J, Wang G, Yin B, Divino CM, Chen SH (2010) Immune stimulatory receptor CD40 is required for T-cell suppression and T regulatory cell activation mediated by myeloid-derived suppressor cells in cancer. Cancer Res 70(1):99–108. doi:10.1158/0008-5472.CAN-09-1882 0008-5472.CAN-09-1882 [pii]

59. Gallina G, Dolcetti L, Serafini P, De Santo C, Marigo I, Colombo MP, Basso G, Brombacher F, Borrello I, Zanovello P, Bicciato S, Bronte V (2006) Tumors induce a subset of inflammatory monocytes with immunosuppressive activity on CD8 + T cells. J Clin Invest 116(10):2777–2790. doi:10.1172/JCI28828

60. Yang R, Cai Z, Zhang Y, Yutzy WHt, Roby KF, Roden RB (2006) CD80 in immune suppression by mouse ovarian carcinoma-associated Gr-1 + CD11b + myeloid cells. Cancer Res 66(13):6807–6815. doi:10.1158/0008-5472.CAN-05-3755 66/13/6807 [pii]

61. Haile LA, Gamrekelashvili J, Manns MP, Korangy F, Greten TF (2010) CD49d is a new marker for distinct myeloid-derived suppressor cell subpopulations in mice. J Immunol 185 (1):203–210. doi:10.4049/jimmunol.0903573

62. Kohanbash G, McKaveney K, Sakaki M, Ueda R, Mintz AH, Amankulor N, Fujita M, Ohlfest JR, Okada H (2013) GM-CSF promotes the immunosuppressive activity of glioma-infiltrating myeloid cells through interleukin-4 receptor-alpha. Cancer Res 73 (21):6413–6423. doi:10.1158/0008-5472.CAN-12-4124

63. Youn JI, Gabrilovich DI (2010) The biology of myeloid-derived suppressor cells: the blessing and the curse of morphological and functional heterogeneity. Eur J Immunol 40 (11):2969–2975. doi:10.1002/eji.201040895

64. Maenhout SK, Van Lint S, Emeagi PU, Thielemans K, Aerts JL (2014) Enhanced suppressive capacity of tumor-infiltrating myeloid-derived suppressor cells compared with their peripheral counterparts. Int J Cancer 134(5):1077–1090. doi:10.1002/ijc.28449

65. Ostrand-Rosenberg S, Sinha P (2009) Myeloid-derived suppressor cells: linking inflammation and cancer. J Immunol 182(8):4499–4506. doi:10.4049/jimmunol.0802740

66. Vasquez-Dunddel D, Pan F, Zeng Q, Gorbounov M, Albesiano E, Fu J, Blosser RL, Tam AJ, Bruno T, Zhang H, Pardoll D, Kim Y (2013) STAT3 regulates arginase-I in myeloid-derived suppressor cells from cancer patients. J Clin Invest 123(4):1580–1589. doi:10.1172/JCI60083

67. Bronte V, Zanovello P (2005) Regulation of immune responses by L-arginine metabolism. Nat Rev Immunol 5(8):641–654. doi:10.1038/nri1668

68. Albeituni SH, Ding C, Yan J (2013) Hampering immune suppressors: therapeutic targeting of myeloid-derived suppressor cells in cancer. Cancer J 19(6):490–501. doi:10.1097/PPO.0000000000000006 00130404-201311000-00006 [pii]

69. Wang L, Chang EW, Wong SC, Ong SM, Chong DQ, Ling KL (2013) Increased myeloid-derived suppressor cells in gastric cancer correlate with cancer stage and plasma S100A8/A9 proinflammatory proteins. J Immunol 190(2):794–804. doi:10.4049/jimmunol.1202088

70. Zea AH, Rodriguez PC, Atkins MB, Hernandez C, Signoretti S, Zabaleta J, McDermott D, Quiceno D, Youmans A, O'Neill A, Mier J, Ochoa AC (2005) Arginase-producing myeloid suppressor cells in renal cell carcinoma patients: a mechanism of tumor evasion. Cancer Res 65(8):3044–3048. doi:10.1158/0008-5472.CAN-04-4505 65/8/3044 [pii]

71. Nagaraj S, Schrum AG, Cho HI, Celis E, Gabrilovich DI (2010) Mechanism of T cell tolerance induced by myeloid-derived suppressor cells. J Immunol 184(6):3106–3116. doi:10.4049/jimmunol.0902661 jimmunol.0902661 [pii]

72. Rodriguez PC, Quiceno DG, Zabaleta J, Ortiz B, Zea AH, Piazuelo MB, Delgado A, Correa P, Brayer J, Sotomayor EM, Antonia S, Ochoa JB, Ochoa AC (2004) Arginase I production in the tumor microenvironment by mature myeloid cells inhibits T-cell receptor expression and antigen-specific T-cell responses. Cancer Res 64(16):5839–5849. doi:10.1158/0008-5472.CAN-04-0465 64/16/5839 [pii]

73. Serafini P, Mgebroff S, Noonan K, Borrello I (2008) Myeloid-derived suppressor cells promote cross-tolerance in B-cell lymphoma by expanding regulatory T cells. Cancer Res 68 (13):5439–5449. doi:10.1158/0008-5472.CAN-07-6621 68/13/5439 [pii]
74. Filipazzi P, Valenti R, Huber V, Pilla L, Canese P, Iero M, Castelli C, Mariani L, Parmiani G, Rivoltini L (2007) Identification of a new subset of myeloid suppressor cells in peripheral blood of melanoma patients with modulation by a granulocyte-macrophage colony-stimulation factor-based antitumor vaccine. J Clin Oncol 25(18):2546–2553. doi:10.1200/JCO.2006.08.5829
75. Abad C, Nobuta H, Li J, Kasai A, Yong WH, Waschek JA (2014) Targeted STAT3 disruption in myeloid cells alters immunosuppressor cell abundance in a murine model of spontaneous medulloblastoma. J Leukoc Biol 95(2):357–367. doi:10.1189/jlb.1012531 jlb. 1012531[pii]
76. Kortylewski M, Kujawski M, Wang T, Wei S, Zhang S, Pilon-Thomas S, Niu G, Kay H, Mule J, Kerr WG, Jove R, Pardoll D, Yu H (2005) Inhibiting Stat3 signaling in the hematopoietic system elicits multicomponent antitumor immunity. Nat Med 11(12):1314–1321. doi:10.1038/ nm1325 nm1325 [pii]
77. Kinjyo I, Inoue H, Hamano S, Fukuyama S, Yoshimura T, Koga K, Takaki H, Himeno K, Takaesu G, Kobayashi T, Yoshimura A (2006) Loss of SOCS3 in T helper cells resulted in reduced immune responses and hyperproduction of interleukin 10 and transforming growth factor-beta 1. J Exp Med 203(4):1021–1031. doi:10.1084/jem.20052333 jem.20052333 [pii]
78. Yu J, Wang Y, Yan F, Li H, Ren X (2013) Response to comment on "Myeloid-derived suppressor cells suppress antitumor immune responses through IDO expression and correlate with lymph node metastasis in patients with breast cancer". J Immunol 190(11):5341–5342. doi:10.4049/jimmunol.1390024 190/11/5341-a [pii]
79. Curti A, Pandolfi S, Valzasina B, Aluigi M, Isidori A, Ferri E, Salvestrini V, Bonanno G, Rutella S, Durelli I, Horenstein AL, Fiore F, Massaia M, Colombo MP, Baccarani M, Lemoli RM (2007) Modulation of tryptophan catabolism by human leukemic cells results in the conversion of CD25- into CD25 + T regulatory cells. Blood 109(7):2871–2877. doi:10. 1182/blood-2006-07-036863 blood-2006-07-036863 [pii]
80. Yu J, Sun J, Wang SE, Li H, Cao S, Cong Y, Liu J, Ren X (2011) Upregulated expression of indoleamine 2, 3-dioxygenase in primary breast cancer correlates with increase of infiltrated regulatory T cells in situ and lymph node metastasis. Clin Dev Immunol 2011:469135. doi:10.1155/2011/469135
81. Corzo CA, Cotter MJ, Cheng P, Cheng F, Kusmartsev S, Sotomayor E, Padhya T, McCaffrey TV, McCaffrey JC, Gabrilovich DI (2009) Mechanism regulating reactive oxygen species in tumor-induced myeloid-derived suppressor cells. J Immunol 182(9):5693–5701
82. Bronte V, Serafini P, De Santo C, Marigo I, Tosello V, Mazzoni A, Segal DM, Staib C, Lowel M, Sutter G, Colombo MP, Zanovello P (2003) IL-4-induced arginase 1 suppresses alloreactive T cells in tumor-bearing mice. J Immunol 170(1):270–278
83. Xia Y, Roman LJ, Masters BS, Zweier JL (1998) Inducible nitric-oxide synthase generates superoxide from the reductase domain. J Biol Chem 273(35):22635–22639
84. Kusmartsev S, Nefedova Y, Yoder D, Gabrilovich DI (2004) Antigen-specific inhibition of CD8 + T cell response by immature myeloid cells in cancer is mediated by reactive oxygen species. J Immunol 172(2):989–999
85. Nefedova Y, Fishman M, Sherman S, Wang X, Beg AA, Gabrilovich DI (2007) Mechanism of all-trans retinoic acid effect on tumor-associated myeloid-derived suppressor cells. Cancer Res 67(22):11021–11028
86. Ando T, Mimura K, Johansson CC, Hanson MG, Mougiakakos D, Larsson C, Martins da Palma T, Sakurai D, Norell H, Li M, Nishimura MI, Kiessling R (2008) Transduction with the antioxidant enzyme catalase protects human T cells against oxidative stress. J Immunol 181(12):8382–8390 181/12/8382 [pii]
87. Hildeman DA, Mitchell T, Aronow B, Wojciechowski S, Kappler J, Marrack P (2003) Control of Bcl-2 expression by reactive oxygen species. Proc Natl Acad Sci USA 100 (25):15035–15040. doi:10.1073/pnas.1936213100 1936213100 [pii]

88. Kujawski M, Kortylewski M, Lee H, Herrmann A, Kay H, Yu H (2008) Stat3 mediates myeloid cell-dependent tumor angiogenesis in mice. J Clin Invest 118(10):3367–3377. doi:10.1172/JCI35213

89. Yang L, Huang J, Ren X, Gorska AE, Chytil A, Aakre M, Carbone DP, Matrisian LM, Richmond A, Lin PC, Moses HL (2008) Abrogation of TGF beta signaling in mammary carcinomas recruits Gr-1 + CD11b + myeloid cells that promote metastasis. Cancer Cell 13 (1):23–35. doi:10.1016/j.ccr.2007.12.004 S1535-6108(07)00373-X [pii]

90. Ahn GO, Brown JM (2008) Matrix metalloproteinase-9 is required for tumor vasculogenesis but not for angiogenesis: role of bone marrow-derived myelomonocytic cells. Cancer Cell 13 (3):193–205. doi:10.1016/j.ccr.2007.11.032 S1535-6108(08)00002-0 [pii]

91. Zhang W, Zhang C, Li W, Deng J, Herrmann A, Priceman SJ, Liang W, Shen S, Pal SK, Hoon DS, Yu H (2015) CD8 + T-cell immunosurveillance constrains lymphoid premetastatic myeloid cell accumulation. Eur J Immunol 45(1):71–81. doi:10.1002/eji.201444467

92. Auphan-Anezin N, Schmitt-Verhulst AM (2015) Silence STAT3 in the procancer niche… and activate CD8 + T cells to kill premetastatic myeloid intruders. Eur J Immunol 45(1):44–48. doi:10.1002/eji.201445300

93. Hiratsuka S, Watanabe A, Aburatani H, Maru Y (2006) Tumour-mediated upregulation of chemoattractants and recruitment of myeloid cells predetermines lung metastasis. Nat Cell Biol 8(12):1369–1375. doi:10.1038/ncb1507 ncb1507 [pii]

94. Kaplan RN, Rafii S, Lyden D (2006) Preparing the "soil": the premetastatic niche. Cancer Res 66(23):11089–11093. doi:10.1158/0008-5472.CAN-06-2407 66/23/11089 [pii]

95. Kaplan RN, Riba RD, Zacharoulis S, Bramley AH, Vincent L, Costa C, MacDonald DD, Jin DK, Shido K, Kerns SA, Zhu Z, Hicklin D, Wu Y, Port JL, Altorki N, Port ER, Ruggero D, Shmelkov SV, Jensen KK, Rafii S, Lyden D (2005) VEGFR1-positive haematopoietic bone marrow progenitors initiate the pre-metastatic niche. Nature 438 (7069):820–827. doi:10.1038/nature04186 nature04186 [pii]

96. Yan HH, Pickup M, Pang Y, Gorska AE, Li Z, Chytil A, Geng Y, Gray JW, Moses HL, Yang L (2010) Gr-1 + CD11b + myeloid cells tip the balance of immune protection to tumor promotion in the premetastatic lung. Cancer Res 70(15):6139–6149. doi:10.1158/0008-5472. CAN-10-0706 0008-5472.CAN-10-0706 [pii]

97. Diaz-Montero CM, Salem ML, Nishimura MI, Garrett-Mayer E, Cole DJ, Montero AJ (2009) Increased circulating myeloid-derived suppressor cells correlate with clinical cancer stage, metastatic tumor burden, and doxorubicin-cyclophosphamide chemotherapy. Cancer Immunol Immunother 58(1):49–59. doi:10.1007/s00262-008-0523-4

98. Draghiciu O, Lubbers J, Nijman HW, Daemen T (2015) Myeloid derived suppressor cells-An overview of combat strategies to increase immunotherapy efficacy. Oncoimmunology 4(1): e954829. doi:10.4161/21624011.2014.954829 954829 [pii]

99. Bill MA, Fuchs JR, Li C, Yui J, Bakan C, Benson DM Jr, Schwartz EB, Abdelhamid D, Lin J, Hoyt DG, Fossey SL, Young GS, Carson WE 3rd, Li PK, Lesinski GB (2010) The small molecule curcumin analog FLLL32 induces apoptosis in melanoma cells via STAT3 inhibition and retains the cellular response to cytokines with anti-tumor activity. Mol Cancer 9:165. doi:10.1186/1476-4598-9-165 1476-4598-9-165 [pii]

100. Lu P, Yu B, Xu J (2012) Cucurbitacin B regulates immature myeloid cell differentiation and enhances antitumor immunity in patients with lung cancer. Cancer Biother Radiopharm 27 (8):495–503. doi:10.1089/cbr.2012.1219

101. Tu SP, Jin H, Shi JD, Zhu LM, Suo Y, Lu G, Liu A, Wang TC, Yang CS (2012) Curcumin induces the differentiation of myeloid-derived suppressor cells and inhibits their interaction with cancer cells and related tumor growth. Cancer Prev Res (Phila) 5(2):205–215. doi:10. 1158/1940-6207.CAPR-11-0247 1940-6207.CAPR-11-0247 [pii]

102. Lin L, Deangelis S, Foust E, Fuchs J, Li C, Li PK, Schwartz EB, Lesinski GB, Benson D, Lu J, Hoyt D, Lin J (2010) A novel small molecule inhibits STAT3 phosphorylation and DNA binding activity and exhibits potent growth suppressive activity in human cancer cells. Mol Cancer 9:217. doi:10.1186/1476-4598-9-217 1476-4598-9-217 [pii]

103. Zhou J, Wu J, Chen X, Fortenbery N, Eksioglu E, Kodumudi KN, Pk EB, Dong J, Djeu JY, Wei S (2011) Icariin and its derivative, ICT, exert anti-inflammatory, anti-tumor effects, and modulate myeloid derived suppressive cells (MDSCs) functions. Int Immunopharmacol 11 (7):890–898. doi:10.1016/j.intimp.2011.01.007 S1567-5769(11)00027-0 [pii]

104. Ko JS, Rayman P, Ireland J, Swaidani S, Li G, Bunting KD, Rini B, Finke JH, Cohen PA (2010) Direct and differential suppression of myeloid-derived suppressor cell subsets by sunitinib is compartmentally constrained. Cancer Res 70(9):3526–3536. doi:10.1158/0008-5472.CAN-09-3278 0008-5472.CAN-09-3278 [pii]

105. Ko JS, Zea AH, Rini BI, Ireland JL, Elson P, Cohen P, Golshayan A, Rayman PA, Wood L, Garcia J, Dreicer R, Bukowski R, Finke JH (2009) Sunitinib mediates reversal of myeloid-derived suppressor cell accumulation in renal cell carcinoma patients. Clin Cancer Res 15(6):2148–2157. doi:10.1158/1078-0432.CCR-08-1332 1078-0432.CCR-08-1332 [pii]

106. Kodera Y, Katanasaka Y, Kitamura Y, Tsuda H, Nishio K, Tamura T, Koizumi F (2011) Sunitinib inhibits lymphatic endothelial cell functions and lymph node metastasis in a breast cancer model through inhibition of vascular endothelial growth factor receptor 3. Breast Cancer Res 13(3):R66. doi:10.1186/bcr2903 bcr2903 [pii]

107. Finke JH, Rini B, Ireland J, Rayman P, Richmond A, Golshayan A, Wood L, Elson P, Garcia J, Dreicer R, Bukowski R (2008) Sunitinib reverses type-1 immune suppression and decreases T-regulatory cells in renal cell carcinoma patients. Clin Cancer Res 14(20):6674–6682

108. Holash J, Davis S, Papadopoulos N, Croll SD, Ho L, Russell M, Boland P, Leidich R, Hylton D, Burova E, Ioffe E, Huang T, Radziejewski C, Bailey K, Fandl JP, Daly T, Wiegand SJ, Yancopoulos GD, Rudge JS (2002) VEGF-Trap: a VEGF blocker with potent antitumor effects. Proc Natl Acad Sci USA 99(17):11393–11398. doi:10.1073/pnas.172398299 172398299 [pii]

109. Fricke I, Mirza N, Dupont J, Lockhart C, Jackson A, Lee JH, Sosman JA, Gabrilovich DI (2007) Vascular endothelial growth factor-trap overcomes defects in dendritic cell differentiation but does not improve antigen-specific immune responses. Clin Cancer Res 13(16):4840–4848

110. Kusmartsev S, Eruslanov E, Kubler H, Tseng T, Sakai Y, Su Z, Kaliberov S, Heiser A, Rosser C, Dahm P, Siemann D, Vieweg J (2008) Oxidative stress regulates expression of VEGFR1 in myeloid cells: link to tumor-induced immune suppression in renal cell carcinoma. J Immunol 181(1):346–353 181/1/346 [pii]

111. Rodriguez PC, Ernstoff MS, Hernandez C, Atkins M, Zabaleta J, Sierra R, Ochoa AC (2009) Arginase I-producing myeloid-derived suppressor cells in renal cell carcinoma are a subpopulation of activated granulocytes. Cancer Res 69(4):1553–1560. doi:10.1158/0008-5472.CAN-08-1921

112. Sumida K, Wakita D, Narita Y, Masuko K, Terada S, Watanabe K, Satoh T, Kitamura H, Nishimura T (2012) Anti-IL-6 receptor mAb eliminates myeloid-derived suppressor cells and inhibits tumor growth by enhancing T-cell responses. Eur J Immunol 42(8):2060–2072. doi:10.1002/eji.201142335

113. Liby KT, Yore MM, Sporn MB (2007) Triterpenoids and rexinoids as multifunctional agents for the prevention and treatment of cancer. Nat Rev Cancer 7(5):357–369. doi:10.1038/nrc2129 nrc2129 [pii]

114. Konopleva M, Zhang W, Shi YX, McQueen T, Tsao T, Abdelrahim M, Munsell MF, Johansen M, Yu D, Madden T, Safe SH, Hung MC, Andreeff M (2006) Synthetic triterpenoid 2-cyano-3,12-dioxooleana-1,9-dien-28-oic acid induces growth arrest in HER2-overexpressing breast cancer cells. Mol Cancer Ther 5(2):317–328. doi:10.1158/1535-7163.MCT-05-0350 5/2/317 [pii]

115. Ahmad R, Raina D, Meyer C, Kufe D (2008) Triterpenoid CDDO-methyl ester inhibits the Janus-activated kinase-1 (JAK1) signal transducer and activator of transcription-3 (STAT3) pathway by direct inhibition of JAK1 and STAT3. Cancer Res 68(8):2920–2926. doi:10.1158/0008-5472.CAN-07-3036 68/8/2920 [pii]

116. Ozao-Choy J, Ma G, Kao J, Wang GX, Meseck M, Sung M, Schwartz M, Divino CM, Pan PY, Chen SH (2009) The novel role of tyrosine kinase inhibitor in the reversal of immune suppression and modulation of tumor microenvironment for immune-based cancer therapies. Cancer Res 69(6):2514–2522. doi:10.1158/0008-5472.CAN-08-4709 0008-5472. CAN-08-4709 [pii]

117. Kerkar SP, Goldszmid RS, Muranski P, Chinnasamy D, Yu Z, Reger RN, Leonardi AJ, Morgan RA, Wang E, Marincola FM, Trinchieri G, Rosenberg SA, Restifo NP (2011) IL-12 triggers a programmatic change in dysfunctional myeloid-derived cells within mouse tumors. J Clin Invest 121(12):4746–4757

118. Liechtenstein T, Perez-Janices N, Blanco-Luquin I, Goyvaerts C, Schwarze J, Dufait I, Lanna A, Ridder M, Guerrero-Setas D, Breckpot K, Escors D (2014) Anti-melanoma vaccines engineered to simultaneously modulate cytokine priming and silence PD-L1 characterized using myeloid-derived suppressor cells as a readout of therapeutic efficacy. Oncoimmunology 3(7):e945378. doi:10.4161/21624011.2014.945378 945378 [pii]

119. Zoglmeier C, Bauer H, Norenberg D, Wedekind G, Bittner P, Sandholzer N, Rapp M, Anz D, Endres S, Bourquin C (2011) CpG blocks immunosuppression by myeloid-derived suppressor cells in tumor-bearing mice. Clin Cancer Res 17(7):1765–1775. doi:10.1158/ 1078-0432.CCR-10-2672 1078-0432.CCR-10-2672 [pii]

120. Shirota Y, Shirota H, Klinman DM (2012) Intratumoral injection of CpG oligonucleotides induces the differentiation and reduces the immunosuppressive activity of myeloid-derived suppressor cells. J Immunol 188(4):1592–1599. doi:10.4049/jimmunol.1101304 jimmunol. 1101304 [pii]

121. Kortylewski M, Swiderski P, Herrmann A, Wang L, Kowolik C, Kujawski M, Lee H, Scuto A, Liu Y, Yang C, Deng J, Soifer HS, Raubitschek A, Forman S, Rossi JJ, Pardoll DM, Jove R, Yu H (2009) In vivo delivery of siRNA to immune cells by conjugation to a TLR9 agonist enhances antitumor immune responses. Nat Biotechnol 27(10):925–932. doi:10.1038/nbt.1564 nbt.1564 [pii]

122. Zhang Q, Hossain DM, Nechaev S, Kozlowska A, Zhang W, Liu Y, Kowolik CM, Swiderski P, Rossi JJ, Forman S, Pal S, Bhatia R, Raubitschek A, Yu H, Kortylewski M (2013) TLR9-mediated siRNA delivery for targeting of normal and malignant human hematopoietic cells in vivo. Blood 121(8):1304–1315. doi:10.1182/blood-2012-07-442590 blood-2012-07-442590 [pii]

123. Hossain DM, Pal SK, Moreira D, Duttagupta P, Zhang Q, Won H, Jones J, D'Apuzzo M, Forman S, Kortylewski M (2015) TLR9-Targeted STAT3 Silencing Abrogates Immunosuppressive Activity of Myeloid-Derived Suppressor Cells from Prostate Cancer Patients. Clin Cancer Res 21(16):3771–3782. doi:10.1158/1078-0432.CCR-14-3145 1078-0432.CCR-14-3145 [pii]

124. Goyvaerts C, De Groeve K, Dingemans J, Van Lint S, Robays L, Heirman C, Reiser J, Zhang XY, Thielemans K, De Baetselier P, Raes G, Breckpot K (2012) Development of the Nanobody display technology to target lentiviral vectors to antigen-presenting cells. Gene Ther 19(12):1133–1140. doi:10.1038/gt.2011.206

125. Goyvaerts C, Dingemans J, De Groeve K, Heirman C, Van Gulck E, Vanham G, De Baetselier P, Thielemans K, Raes G, Breckpot K (2013) Targeting of Human Antigen-Presenting Cell Subsets. J Virol 87(20):11304–11308. doi:10.1128/JVI.01498-13

126. Amoozgar Z, Goldberg MS (2015) Targeting myeloid cells using nanoparticles to improve cancer immunotherapy. Adv Drug Deliv Rev 30(91):38–51

Chapter 7
Future Perspectives

David Escors and Grazyna Kochan

Abstract The participation of myeloid cells in tumor progression and metastasis has been known for a long time. The role of M2 macrophages, tolerogenic DCs, and N2 neutrophils in tumor immunology has been researched extensively. About 10 years ago, a "re-discovered" new myeloid player named myeloid-derived suppressor cell (MDSC) was put on the spot. However, its precise origin and nature was a subject of some scientific debate. MDSCs turned out to be highly heterogeneous, especially in humans, and exhibiting cancer type-specific properties and characteristics. And despite all recent advances in MDSC research, many questions remain unanswered. In this chapter we will summarize the main subjects addressed in this book and point out the questions that remain unanswered.

Keywords Inmunosuppression · PDL1 · PD1 · Therapeutic antibodies · Signaling pathways · Kinase inhibitors · Chemotherapy · Immunotherapy

7.1 Myeloid Cells and Cancer

The tumor microenvironment is composed not only by cancerous cells, but also other associated cell types including fibroblasts, endothelial cells, and infiltrating immune cells. Within the tumor, there is a balance between cells with antitumor

D. Escors (✉) · G. Kochan
Navarrabiomed-Biomedical Research Centre, Fundación Miguel Servet, IdiSNA,
Calle Irunlarrea 3, 31008 Pamplona, Navarra, Spain
e-mail: descorsm@navarra.es

G. Kochan
e-mail: grkochan@navarra.es

D. Escors
Immunomodulation Group, Rayne Building University College London,
5 University Street, London WC1E 6JF, UK

© The Author(s) 2016
D. Escors et al., *Myeloid-Derived Suppressor Cells and Cancer*,
SpringerBriefs in Immunology, DOI 10.1007/978-3-319-26821-7_7

capacities and immunosuppressive properties. The outcome is usually favorable for immunosuppressive cells, which also exert strong pro-angiogenic effects and accelerate tumor growth and metastasis.

Do MDSCs comprise a distinct myeloid lineage?

As discussed in the first chapter, the infiltration of tumors by myeloid cells was observed and described before the 70s [1]. In fact, infiltration of tumors with these cells was a sign of poor prognosis rather than proof of antitumor responses. However, cells of the myeloid lineage are quite heterogeneous and include dendritic cells, macrophages, and granulocytes. These cell types can also possess either stimulatory properties or immunosuppressive capacities. Thus, when MDSCs were defined according to the expression of CD11b and GR1 (in mice), there was some opposition in considering them as a lineage apart [2]. Even more, monocytic MDSCs show a phenotype that closely resembles inflammatory monocytes. Granulocytic MDSCs are phenotypically closely similar to neutrophils [3, 4].

Therefore, the main question that remains to be answered is whether MDSCs are truly a lineage apart, they are "alternative" forms of monocytes or granulocytes or they convert into one another [5–7]. Whether they are considered as a *bona fide* myeloid lineage or not, their role in tumor progression is not questioned. Infiltrating myeloid cells are present within the tumor and protect cancer against both conventional and immunotherapies.

What is the relationship between MDSCs and other regulatory cell lineages?

The tumor environment can be conserved as a complex "organ" under chronic inflammatory conditions which favor the infiltration of regulatory cells [8]. These strongly immunosuppressive cells play an important role in tumor biology, as they suppress antitumor immune responses, favor tumor progression, tissue repair and neoangiogenesis. These tumor-promoting functions accelerate cancer metastasis. Immunosuppressive infiltrating cells comprise tumor-associated M2 macrophages, tumor-associated neutrophils, tolerogenic DCs, and regulatory T and B cells. Recently, there has been growing experimental evidence that MDSCs do not function on their own, but cooperate with other tumor-associated regulatory cells. This includes crosstalk with macrophages, induction of regulatory T cells, and with regulatory tumor-associated B cells [9, 10]. Interestingly, all these cell types share many of the suppressive pathways, including TGFβ and IL10 production, consumption of essential amino acids, and cell-to-cell contact dependent immunosuppression [8]. Thus, not only MDSCs cooperate with other tumor-associated cells, but they also share common procarcinogenic mechanisms. The identification of their interactions will surely open new opportunities for therapeutic intervention by simultaneously targeting several of these cell types within the tumor.

7.2 Differentiation of Myeloid-Derived Suppressor Cells

As mentioned above, the specific nature and ontogeny of MDSCs are still under debate, possibly due to their phenotypic plasticity and heterogeneity. Therefore, the study of the MDSC differentiation pathways will help to understand whether MDSCs can be considered a lineage of its own right or just a collection of heterogenous myeloid cells at a various differentiation stages.

Murine MDSC differentiation

Without any doubt, murine systems are usually way ahead of their human counterparts. This is also true for MDSCs, which can be easily obtained from mice by inducing tumor growth in vivo, or by differentiating MDSCs from bone marrow cells in vitro. As exposed in various parts of this book, there is a somewhat "strong" consensus on murine MDSC phenotype [1, 11]. These cells express CD11b and high levels of GR1. Then, according to their pattern of ly6C-Ly6G expression, they can be further classified as monocyte (Ly6C^{high}, Ly6G$^{low/neg}$) or granulocytic (Ly6C$^+$ Ly6G^{high}) [12–14]. Unfortunately, these phenotypes are equivalent to those of inflammatory monocytes and neutrophils, respectively. Thus, at the end only the immunosuppressive properties define them. Recent data has shown that melanoma MDSCs present a kinase signature that controls their suppressive activities [15, 16]. Nevertheless, although these kinase signatures explain the nature of MDSCs at least functionally, all these data does not clarify their ontogeny yet.

In vitro systems have not shed much light on this subject, as it would have been expected. Each system has its advantages and limitations, but so far the MDSC differentiation pathway (if there is a single one) is still poorly understood [11]. Therefore, even though some steps have been undertaken toward the development of efficacious ex vivo MDSC production methods [7, 13–15, 17], the faithful replication of the MDSC differentiation pathways in vitro and in vivo is a pending subject.

Human MDSC differentiation

Compared to murine systems, very little is known about human MDSCs. This is directly caused by the intrinsic difficulties of working with samples from patients with cancer. Most of the studies are centered on peripheral blood cells, and the in vitro MDSC systems are highly inefficient as they do not use fully pluripotent hematopoietic precursors [11]. In addition to these important drawbacks, the human MDSC phenotype is still largely undefined [18, 19]. Some attempts have been made at classifying MDSC types in humans according to phenotype, tumor models, and sources of cells [20]. Thus, in the human system we might have three possibilities. First, it might be intrinsically heterogenous with several types of co-existing MDSCs. Second, there might not be MDSCs at all (as we understand from the murine system) but a collection of myeloid cells at different differentiation stages. Or third, we are studying mainly circulating MDSCs from peripheral blood rather than homogeneous cell populations derived from bone marrow.

Efficient in vitro systems should be developed for human MDSCs, and this will surely help deciding whether human MDSCs are comparable to their murine counterparts. This is also a key issue, as most cancer therapies are tested first in murine systems. Although major advances have been made toward this goal, these human MDSC differentiation systems are still poorly efficient.

7.3 Targeting MDSC-Specific Pathways for Therapy

While there are still many open questions on MDSC biology, in practical terms their elimination from a tumor-bearing subject improves anticancer therapies. Thus, obviating the fact whether these cells comprise a specific myeloid lineage or not, much is being understood on their immunosuppressive mechanisms. This knowledge uncovers opportunities for therapeutic interventions. From early studies, it was observed that MDSCs could exert suppressive activities by secreting factors, or by cell-to-cell contact mechanisms. Apart from the classical secretion of immunosuppressive cytokines such as TGFβ or IL10, similarly to M2 macrophages the arginine metabolism was shown to play a very important role in their activities [7, 21]. Arginine is processed in MDSCs by two enzymes, iNOS and arginase-1. Blocking the activity of both enzymes improves antitumor activities in mouse models [7].

The tumor environment as a target

The tumor environment as a whole is also a therapeutic target. Cancer cells and tumor-infiltrating cells are under a very strong oxidative stress, and upregulate detoxifying enzymes and ROS scavenging proteins. MDSCs have been shown to selectively upregulate the P450 reductase, and this upregulation explains the anti-MDSC properties of Paclitaxel [22]. This chemotherapy drug needs to be activated by P450R to acquire cytotoxic activities. As conventional immunogenic DCs express lower levels of P450R, these cells are by far less sensitive to Paclitaxel than MDSCs [14]. This is just but one example on how to exploit these tumor-induced cellular targets.

MDSC signaling pathways as a target

Interestingly, there is a growing field of research on MDSC signaling, as tyrosine kinase inhibitors and other chemotherapy drugs eliminate MDSCs both in murine models and human patients. Again, pathways shared by cancer cells and tumor-associated cells are also present in MDSCs [14–16]. Therefore, and unknowingly, many of the anticancer drugs that were designed to directly attack cancer cells, also have anti-MDSC properties. Thus, all these shared pathways are susceptible of therapeutic intervention in a straightforwardly manner. As already discussed in a previous chapter, one of those is the STAT3-dependent signaling pathway [23, 24]. This pathway regulates cell growth, survival, and inflammation. It is also activated by IL6, a cytokine known to contribute to MDSC differentiation [25]. STAT3 is constitutively activated in cancer cells, tumor cells (which includes cancerous and associated cells), and in tumor infiltrating cells of the immune system [26].

Extensive work has been performed on STAT3 in macrophages. IL10 is also a potent induction of STAT3, and its phosphorylation in macrophages leads to their polarization toward immunosuppressive subsets [27, 28]. This pathway together with others such as PI3 K/AKT acts as a safeguard against uncontrolled inflammation. However, cancer can turn on this pathway to inhibit antitumor immune responses [29, 30]. MDSCs seem to activate the STAT3 pathway in cancer, and therapeutic strategies devised to act upon tumor-infiltrating macrophages and DCs will probably be successful in counteracting MDSC-suppressive activities.

Apart from STAT3, the implication of several intracellular pathways on MDSC biology has also been described [14, 15]. These pathways are linked to cell survival, anti-inflammatory responses and stress responses against oxidative stress. Thus, other tumor-associated cells also share them. Moreover, there is a specific kinase profile in MDSCs that separates them from other conventional myeloid immunogenic cell types. The PI3 K, AKT, and the SRC family of kinases are highly upregulated in murine melanoma MDSCs and their expression differentiates them from conventional myeloid DCs [15]. MDSCs obtained from other tumor backgrounds, especially breast cancer MDSCs, also show increased levels of AKT, and a gene expression profile characteristic of the activity of SRC family members, particularly HCK and FYN kinases [15]. AKT and PI3 K are also highly activated in MDSCs. Interestingly, tumor-infiltrating melanoma MDSCs specifically activate ERK1 and PKC kinases, which are also known to be activated in tumor cells [8, 15, 31].

There is currently a wide range of small molecules that target these pathways, activated both in cancer cells and MDSCs. The Ras-Raf-MEK-ERK signaling axis is probably one of the most studied for the development of anticancer treatments [32]. MDSC differentiation has been found to be particularly affected by AKT and MEK inhibitors, while their immunogenic myeloid DC counterparts were largely unaffected [15]. Moreover, MEK inhibition enhances DC differentiation and activates DC-mediated antitumor activities [15, 33–35].

Thus, many anti-neoplastic treatments also have "beneficial collateral" effects on MDSCs. The assessment of these anticancer drugs over MDSCs will surely shed light on their multiple mechanisms of action over the immune system [11]. Furthermore, the specific kinase signature found in MDSCs will facilitate the development of efficacious MDSC-targeted therapies that do not affect immunogenic cells such as DCs.

Interfering with negative co-stimulation of T cells as a target

Similarly to DCs and other myeloid cells, MDSCs are also antigen-presenting cells. However, after antigen presentation by MDSCs, T cells get inactivated, suppressed, or differentiate toward regulatory T cells [13, 36]. There are multiple mechanisms by which MDSCs can exert T cell inhibitory effects, and those include secretion of anti-inflammatory cytokines, consumption of essential amino acids, production of NO and use of negative co-stimulation during antigen presentation to T cells [8].

During antigen presentation, the antigen-presenting cell (APC) presents to T cells complexed to major histocompatibility molecules on their surface (Fig. 7.1). These pMHC complexes are recognized and bind to specific T cell receptors (TCRs) present on the surface of CD4 or CD8 T cells. This recognition sends a signal (signal 1) to the T cells. However, this signal is not sufficient to activate a T cell and leads to T-cell anergy instead [37]. Further interactions between these two cells are required within the immunological synapse. These interactions take place between antigenic peptides receptors on the T cells and their respective ligands on the APC. Some of these interactions will lead to T-cell activation while others will dampen T cells. The integration between all these differing interactions provides a second signal during antigen presentation. This signal 2 will determine whether T cells get activated or not, and the extent of T-cell activation. Positive co-stimulation is represented by the classical

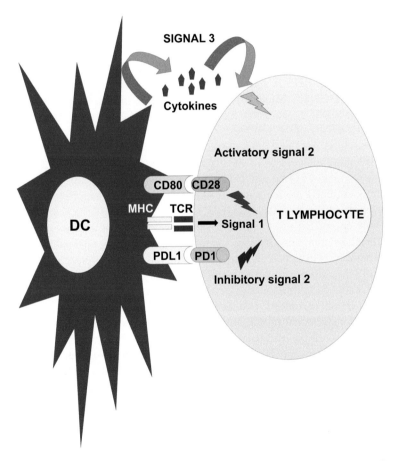

Fig. 7.1 Physiological antigen presentation to T-cells. The scheme represents a DC as an antigen-presenting cell to a T lymphocyte through the MHC-TCR complex as indicated in the immunological synapse between the two cell types. Both positive and negative receptor–ligand interactions take place, as indicated in the picture. These interactions will transmit signals (activatory and inhibitory signal 2, as shown) that together with antigen recognition (signal 1), will regulate T-cell activation or its effector functions. In addition, a third signal is provided within the immunological synapse in the form of secreted cytokines (*top of the figure*). The integration of these three signals within the T cell will determine the level of T-cell activation and its polarization

interaction between CD80 and CD28, on the surfaces of APCs and T cells, respectively. However, there are a high number of interactions that regulate T-cell activation by sending inhibitory signals. For example, CD80-CTLA4 or PDL1-PD1 (Fig. 7.1).

Thus, antibodies which block these interactions have been developed to strengthen T-cell activation by interfering with these interactions. Preventing CTLA4 binding to CD80 has been one of the first to be applied to human therapy and showing success [38]. In fact, MDSCs express very high levels of CD80, which seems to be required for their suppressive functions (Fig. 7.2) [14, 17, 39, 40]. Recently, blocking PDL1-PD1 interactions with antibodies is demonstrating to be a very successful immunotherapy anticancer strategy [41, 42]. While it is widely thought that the mechanism of action takes place at the tumor site by facilitating the attack of the effector T cell, their efficacy in some PDL1-negative tumors indicates that there are other mechanisms of action. In fact, PDL1-PD1 interactions play a key role in antigen presentation. Their interaction following antigen recognition by the T cells facilitates ligand-induced TCR down-modulation while the T cell gets activated and proliferates [33, 43, 44]. TCR expression recovers after one week, and this is a safeguard mechanism that ensures that T cells do not attack their targets until they reach a critical number [45, 46]. Interfering with this interaction leads to hyperproliferative TCR^{high} polyfunctional effector CD8 T cells with strong antitumor activities [33, 47]. Additionally, interference with PDL1 expression also leads to a low level expansion of polyclonal CD8 T cells which probably contribute to anticancer activities [13]. Tumor-infiltrating MDSCs express very high levels of PDL1 (Fig. 7.2) [7, 12–14, 17]. Interference with PDL1 expression on MDSCs converts these cells in T-cell stimulators [48]. It is highly likely that current blocking antibodies used in human therapy are converting MDSCs to efficient immunstimulatory APCs. As there are an increasing number of positive and negative co-stimulatory molecules and antibodies targeting their interaction partners [49, 50], it is highly likely that immunotherapy will become a first-line treatment for cancer. These immunotherapy approaches directly target MDSCs by converting them in immunostimulatory myeloid cells.

Conversion of MDSCs to efficient APCs with antitumor properties

While specific targeting and depletion of MDSCs improves antitumor immune responses [51, 52], an interesting approach that will certainly have a future in anticancer therapies is the conversion of MDSCs into proinflammatory APCs. MDSCs have been shown to possess the potential of differentiation toward other myeloid cell types such as DCs, macrophages, and inflammatory granulocytes. While several cytokines and factors may drive this differentiation, IL12 is proving to be quite efficacious in converting MDSCs to immunogenic myeloid APCs. Thus, direct treatment with IL12 transforms MDSCs into activated antigen-presenting cells [13, 53, 54]. Within the tumor environment, IL12 production leads to a collapse of the tumor stroma, which helps regression and improves antitumor capacities of T cells [55]. It is highly likely that the method of IL12 administration will likely have an impact in its efficacy. So far, local IL12 production within the tumor environment is proving the method of choice as it will surely reduce cytotoxicity from systemic administration.

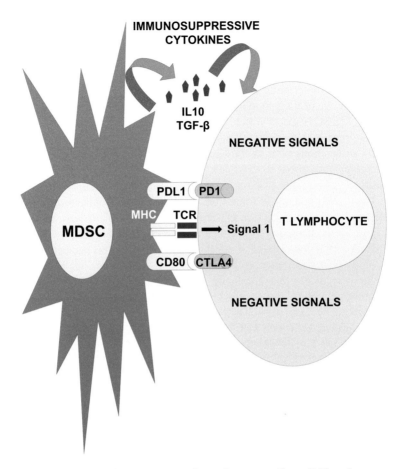

Fig. 7.2 The MDSC as an immunosuppressive antigen-presenting cell. The scheme represents a MDSC presenting antigen to a T lymphocyte through the MHC-TCR complex as indicated in the immunological synapse between the two cell types. Negative receptor–ligand interactions take place primarily when MDSCs present antigen, by upregulating PDL1 binding to PD1 on the T-cell surface, and expressing high levels of CD80 which binds CTLA4 on the T cell, as shown in the figure. These interactions together with antigen recognition (signal 1), inhibits either T-cell activation, or its effector functions. In addition, MDSCs produce high levels of immunosuppressive cytokines, as indicated in the figure. These cytokines will polarize T cells toward tolerogenic subsets such as inducible regulatory T cells (Tregs)

7.4   Summary and Conclusions

Although the participation of myeloid cells on tumor progression and metastasis has been known for a long time, only recently another "subset" of myeloid cells has been added to this picture. This has raised some controversy on their nature and relationship with other myeloid cell types. Nevertheless, whether they represent a *bona fide* myeloid lineage, or another differentiation stage of highly plastic myeloid cells, they strongly possess procarcinogenic properties.

From a scientific point of view, their "true" ontogenetic nature needs to be clarified, especially for human MDSCs. From a practical point of view, tumor-associated myeloid immunosuppressive cells need to be eliminated.

Apart from controversies, their importance in cancer immunology is undeniable. Proof of this is the increasingly higher number of publications dealing with them. An important effort is being devoted to devise efficient differentiation methods for basic research or for cellular therapies. Obtaining MDSCs that resemble tumor-infiltrating subsets is still challenging, although encouraging steps have been recently taken toward this goal in murine systems. The human system is still a pending subject.

Immunotherapy will surely become a first-line anticancer treatment strategy, and MDSCs will surely occupy a central position in anticancer research.

Finally, a clearer view on MDSC biology is emerging from recent research, which highlights the metabolic changes and high differentiation plasticity of the "myeloid cell compartment". However, this plasticity can be used to devise targeted therapies that will eliminate the procarcinogenic myeloid cells and shift differentiation toward immunogenic, protective cells.

Acknowledgments David Escors is funded by a Miguel Servet Fellowship (CP12/03114), a FIS project grant (PI14/00579) from the Instituto de Salud Carlos III, Spain, the Refbio transpyrenaic collaborative project grants (NTBM), a Sandra Ibarra Foundation grant, Gobierno de Navarra Grant (BMED 033-2014), and a Gobierno Vasco BioEf project grant (BIO13/CI/014). Grazyna Kochan is funded by a Caixa Bank Grant.

References

1. Talmadge JE, Gabrilovich DI (2013) History of myeloid-derived suppressor cells. Nat Rev 13(10):739–752
2. Gabrilovich DI, Bronte V, Chen SH, Colombo MP, Ochoa A, Ostrand-Rosenberg S, Schreiber H (2007) The terminology issue for myeloid-derived suppressor cells. Cancer Res 67 (1):425; author reply 426. doi:67/1/425 [pii] 10.1158/0008-5472.CAN-06-3037
3. Fridlender ZG, Sun J, Mishalian I, Singhal S, Cheng G, Kapoor V, Hornq W, Fridlender G, Bayuh R, Worthen GS, Albelda SM (2012) Transcriptomic analysis comparing tumor-associated neutrophils with granulocytic myeloid-derived suppressor cells and normal neutrophils. PLoS ONE 7(2):e31524
4. Youn JI, Collazo M, Shalova IN, Biswas SK, Gabrilovich DI (2012) Characterization of the nature of granulocytic myeloid-derived suppressor cells in tumor-bearing mice. J Leukoc Biol 91(1):167–181. doi:10.1189/jlb.0311177
5. Koffel R, Meshcheryakova A, Warszawska J, Henning A, Wagner K, Jorgl A, Gubi D, Moser D, Hladik A, Hoffmann U, Fischer MB, van der Berg W, Koenders M, Scheinecker C, Gesslbauer B, Knapp S, Strobl H (2014) Monocytic cell differentiation from band-stage neutrophils under inflammatory conditions via MKK6 activation. Blood 124(17):2713–2724
6. Youn JI, Kumar V, Collazo M, Nefedova Y, Condamine T, Cheng P, Villagra A, Antonia S, McCaffrey JC, Fishman M, Sarnaik A, Horna P, Sotomayor E, Gabrilovich DI (2013) Epigenetic silencing of retinoblastoma gene regulates pathologic differentiation of myeloid cells in cancer. Nat Immunol 14(3):211–220. doi:10.1038/ni.2526ni 2526[pii]
7. Dufait I, Schwarze JK, Liechtenstein T, Leonard W, Jiang H, Law K, Verovski V, Escors D, De Ridder M, Breckpot K (2015) Ex vivo generation of myeloid-derived suppressor cells that

model the tumor immunosuppressive environment in colorectal cancer. Oncotarget 6 (14):12369–12382

8. Escors D (2014) Tumour immunogenicity, antigen presentation and immunological barriers in cancer immunotherapy. New J Sci 2014. doi:10.1155/2014/734515 734515[pii]

9. Ostrand-Rosenberg S, Sinha P, Beury DW, Clements VK (2012) Cross-talk between myeloid-derived suppressor cells (MDSC), macrophages, and dendritic cells enhances tumor-induced immune suppression. Semin Cancer Biol 22(4):275–281. doi:10.1016/j.semcancer.2012.01.011

10. Bodogai M, Moritoh K, Lee-Chang C, Hollander CM, Sherman-Baust CA, Wersto RP, Araki Y, Miyoshi I, Yang L, Trinchieri G, Biragyn A (2015) Immune suppressive and pro-metastatic functions of myeloid-derived suppressive cells rely upon education from tumor-associated B cells. Cancer Res. doi:10.1158/0008-5472.CAN-14-3077

11. Escors D, Liechtenstein T, Perez-Janices N, Schwarze J, Dufait I, Goyvaerts C, Lanna A, Arce F, Blanco-Luquin I, Kochan G, Guerrero-Setas D, Breckpot K (2013) Assessing T-cell responses in anticancer immunohterapy: dendritic cells or myeloid-derived suppressor cells? Oncoimmunology 12(10):e26148

12. Youn JI, Nagaraj S, Collazo M, Gabrilovich DI (2008) Subsets of myeloid-derived suppressor cells in tumor-bearing mice. J Immunol 181(8):5791–5802

13. Liechtenstein T, Perez-Janices N, Blanco-Luquin I, Schwarze J, Dufait I, Lanna A, De Ridder M, Guerrero-Setas D, Breckpot K, Escors D (2014) Anti-melanoma vaccines engineered to simultaneously modulate cytokine priming and silence PD-L1 characterized using ex vivo myeloid-derived suppressor cells as a readout of therapeutic efficacy. Oncoimmunology 3:e29178

14. Liechtenstein T, Perez-Janices N, Gato M, Caliendo F, Kochan G, Blanco-Luquin I, Van der Jeught K, Arce F, Guerrero-Setas D, Fernandez-Irigoyen J, Santamaria E, Breckpot K, Escors D (2014) A highly efficient tumor-infiltrating MDSC differentiation system for discovery of anti-neoplastic targets, which circumvents the need for tumor establishment in mice. Oncotarget 5(17):7843–7857

15. Gato-Cañas M, Martinez de Morentin X, Blanco-Luquin I, Fernandez-Irigoyen J, Zudaire I, Liechtenstein T, Arasanz H, Lozano T, Casares N, Knapp S, Chaikuad A, Guerrero-Setas D, Escors D, Kochan G, Santamaria E (2015) A core of kinase-regulated interactomes defines the neoplastic MDSC lineage. Oncotarget In press

16. Aliper AM, Frieden-Korovkina VP, Buzdin A, Roumiantsev SA, Zhavoronkov A (2014) Interactome analysis of myeloid-derived suppressor cells in murine models of colon and breast cancer. Oncotarget 5(22):11345–11353

17. Van der Jeught K, Joe PT, Bialkowski L, Heirman C, Daszkiewicz L, Liechtenstein T, Escors D, Thielemans K, Breckpot K (2014) Intratumoral administration of mRNA encoding a fusokine consisting of IFN-beta and the ectodomain of the TGF-beta receptor II potentiates antitumor immunity. Oncotarget 5(20):10100–10113

18. Peranzoni E, Zilio S, Marigo I, Dolcetti L, Zanovello P, Mandruzzato S, Bronte V (2010) Myeloid-derived suppressor cell heterogeneity and subset definition. Curr Opin Immunol 22(2):238–244. doi:10.1016/j.coi.2010.01.021 S0952-7915(10)00022-1 [pii]

19. Damuzzo V, Pinton L, Desantis G, Solito S, Marigo I, Bronte V, Mandruzzato S (2015) Complexity and challenges in defining myeloid-derived suppressor cells. Cytometry Part B, Clinical cytometry 88(2):77–91. doi:10.1002/cyto.b.21206

20. Solito S, Marigo I, Pinton L, Damuzzo V, Mandruzzato S, Bronte V (2014) Myeloid-derived suppressor cell heterogeneity in human cancers. Ann N Y Acad Sci 1319:47–65. doi:10.1111/nyas.12469

21. Nagaraj S, Collazo M, Corzo CA, Youn JI, Ortiz M, Quiceno D, Gabrilovich DI (2009) Regulatory myeloid suppressor cells in health and disease. Cancer Res 69(19):7503–7506

22. Sevko A, Michels T, Vrohlings M, Umansky L, Beckhove P, Kato M, Shurin GV, Shurin MR, Umansky V (2013) Antitumor effect of paclitaxel is mediated by inhibition of myeloid-derived suppressor cells and chronic inflammation in the spontaneous melanoma model. J Immunol 190(5):2464–2471. doi:10.4049/jimmunol.1202781

23. Emeagi PU, Maenhout S, Dang N, Heirman C, Thielemans K, Breckpot K (2013) Downregulation of Stat3 in melanoma: reprogramming the immune microenvironment as an anticancer therapeutic strategy. Gene Ther. 20(11):1085–1092. doi:10.1038/gt.2013.35 gt201335 [pii]

24. Waight JD, Netherby C, Hensen ML, Miller A, Hu Q, Liu S, Bogner PN, Farren MR, Lee KP, Liu K, Abrams SI (2013) Myeloid-derived suppressor cell development is regulated by a STAT/IRF-8 axis. J Clin Invest 123(10):4464–4478

25. Park SJ, Nakagawa T, Kitamura H, Atsumi T, Kamon H, Sawa S, Kamimura D, Ueda N, Iwakura Y, Ishihara K, Murakami M, Hirano T (2004) IL-6 regulates in vivo dendritic cell differentiation through STAT3 activation. J Immunol 173(6):3844–3854

26. Nefedova Y, Huang M, Kusmartsev S, Bhattacharya R, Cheng P, Salup R, Jove R, Gabrilovich D (2004) Hyperactivation of STAT3 is involved in abnormal differentiation of dendritic cells in cancer. J Immunol 172(1):464–474

27. Niemand C, Nimmesgern A, Haan S, Fischer P, Schaper F, Rossaint R, Heinrich PC, Muller-Newen G (2003) Activation of STAT3 by IL-6 and IL-10 in primary human macrophages is differentially modulated by suppressor of cytokine signaling 3. J Immunol 170(6):3263–3272

28. O'Farrell AM, Liu Y, Moore KW, Mui AL (1998) IL-10 inhibits macrophage activation and proliferation by distinct signaling mechanisms: evidence for Stat3-dependent and -independent pathways. EMBO J 17(4):1006–1018. doi:10.1093/emboj/17.4.1006

29. Yang J, Liao D, Chen C, Liu Y, Chuang TH, Xiang R, Markowitz D, Reisfeld RA, Luo Y (2013) Tumor-associated macrophages regulate murine breast cancer stem cells through a novel paracrine EGFR/Stat3/Sox-2 signaling pathway. Stem cells (Dayton, Ohio) 31(2):248–258. doi:10.1002/stem.1281

30. Wang T, Niu G, Kortylewski M, Burdelya L, Shain K, Zhang S, Bhattacharya R, Gabrilovich D, Heller R, Coppola D, Dalton W, Jove R, Pardoll D, Yu H (2004) Regulation of the innate and adaptive immune responses by Stat-3 signaling in tumor cells. Nat Med 10(1):48–54

31. Arce F, Kochan G, Breckpot K, Stephenson H, Escors D (2012) Selective Activation of Intracellular Signalling Pathways In Dendritic Cells For Cancer Immunotherapy. Anti-Cancer Agents Med Chem 1:29–39

32. Samatar AA, Poulikakos PI (2014) Targeting RAS-ERK signalling in cancer: promises and challenges. Nat Rev Drug Discovery 13(12):928–942. doi:10.1038/nrd4281

33. Karwacz K, Bricogne C, Macdonald D, Arce F, Bennett CL, Collins M, Escors D (2011) PD-L1 co-stimulation contributes to ligand-induced T cell receptor down-modulation on CD8 (+) T cells. EMBO Mol Med 3(10):581–592. doi:10.1002/emmm.201100165

34. Arce F, Breckpot K, Stephenson H, Karwacz K, Ehrenstein MR, Collins M, Escors D (2011) Selective ERK activation differentiates mouse and human tolerogenic dendritic cells, expands antigen-specific regulatory T cells, and suppresses experimental inflammatory arthritis. Arthritis Rheum 63:84–95

35. Escors D, Lopes L, Lin R, Hiscott J, Akira S, Davis RJ, Collins MK (2008) Targeting dendritic cell signalling to regulate the response to immunisation. Blood 111(6):3050–3061. doi:10.1182/blood-2007-11-122408 blood-2007-11-122408 [pii]

36. Luan Y, Mosheir E, Menon MC, Wilson D, Woytovich C, Ochando J, Murphy B (2013) Monocytic myeloid-derived suppressor cells accumulate in renal transplant patients and mediate CD4(+) Foxp3(+) Treg expansion. Am J Transplant 13(12):3123–3131. doi:10.1111/ajt.12461

37. Yamamoto T, Hattori M, Yoshida T (2007) Induction of T-cell activation or anergy determined by the combination of intensity and duration of T-cell receptor stimulation, and sequential induction in an individual cell. Immunology 121(3):383–391. doi:10.1111/j.1365-2567.2007.02586.x

38. Wolchok JD, Hodi FS, Weber JS, Allison JP, Urba WJ, Robert C, O'Day SJ, Hoos A, Humphrey R, Berman DM, Lonberg N, Korman AJ (2013) Development of ipilimumab: a novel immunotherapeutic approach for the treatment of advanced melanoma. Ann N Y Acad Sci 1291(1):1–13. doi:10.1111/nyas.12180

39. Dilek N, Vuillefroy de Silly R, Blancho G, Vanhove B (2012) Myeloid-derived suppressor cells: mechanisms of action and recent advances in their role in transplant tolerance. Front Immunol 3:208. doi:10.3389/fimmu.2012.00208

40. Maenhout SK, Van Lint S, Emeagi PU, Thielemans K, Aerts JL (2014) Enhanced suppressive capacity of tumor-infiltrating myeloid-derived suppressor cells compared to their peripheral counterparts. Int J Cancer 134(5):1077–1090. doi:10.1002/ijc.28449

41. Brahmer JR, Tykodi SS, Chow LQ, Hwu WJ, Topalian SL, Hwu P, Drake CG, Camacho LH, Kauh J, Odunsi K, Pitot HC, Hamid O, Bhatia S, Martins R, Eaton K, Chen S, Salay TM, Alaparthy S, Grosso JF, Korman AJ, Parker SM, Agrawal S, Goldberg SM, Pardoll DM, Gupta A, Wigginton JM (2012) Safety and activity of anti-PD-L1 antibody in patients with advanced cancer. N Engl J Med 366(26):2455–2465

42. Topalian SL, Hodi FS, Brahmer JR, Gettinger SN, Smith DC, McDermott DF, Powderly JD, Carvajal RD, Sosman JA, Atkins MB, Leming PD, Spigel DR, Antonia SJ, Horn L, Drake CG, Pardoll DM, Chen L, Sharfman WH, Anders RA, Taube JM, McMiller TL, Xu H, Korman AJ, Jure-Kunkel M, Agrawal S, McDonald D, Kollia GD, Gupta A, Wigginton JM, Sznol M (2012) Safety, activity, and immune correlates of anti-PD-1 antibody in cancer. N Engl J Med 366(26):2443–2454

43. Karwacz K, Arce F, Bricogne C, Kochan G, Escors D (2012) PD-L1 co-stimulation, ligand-induced TCR down-modulation and anti-tumor immunotherapy. Oncoimmunology 1(1):86–88

44. Escors D, Bricogne C, Arce F, Kochan G, Karwacz K (2012) On the mechanism of T cell receptor down-modulation and its physiological significance. J biosci med 1(1). 2011.5 [pii]

45. Liechtenstein T, Dufait I, Bricogne C, Ianna A, Pen J, Breckpot K, Escors D (2012) PD-L1/PD-1 co-stimulation, a brake for T cell activation and a T cell differentiation signal. J Clin Cell Immunol S12(006):6. doi:10.4172/2155-9899.S12-006

46. Bricogne C, Laranga R, Padella A, Dufait I, Liechtenstein T, Breckpot K, Kochan G, Escors D (2012) Critical Mass Hypothesis of T-Cell Responses and its Application for the Treatment of T-Cell Lymphoma. In: Harvey WK, Jacobs RM (eds) Hodgkin's and T-cell lymphoma: Diagnosis. Nova Publishers, Treatment Options and Prognosis

47. Pen JJ, Keersmaecker BD, Heirman C, Corthals J, Liechtenstein T, Escors D, Thielemans K, Breckpot K (2013) Interference with PD-L1/PD-1 co-stimulation during antigen presentation enhances the multifunctionality of antigen-specific T cells. Gene Ther 21(3):262–271

48. Noman MZ, Desantis G, Janji B, Hasmim M, Karray S, Dessen P, Bronte V, Chouaib S (2014) PD-L1 is a novel direct target of HIF-1alpha, and its blockade under hypoxia enhanced MDSC-mediated T cell activation. J Exp Med 211(5):781–790. doi:10.1084/jem.20131916

49. Melero I, Hervas-Stubbs S, Glennie M, Pardoll DM, Chen L (2007) Immunostimulatory monoclonal antibodies for cancer therapy. Nat Rev 7(2):95–106. doi:10.1038/nrc2051

50. Chen L, Flies DB (2013) Molecular mechanisms of T cell co-stimulation and co-inhibition. Nat Rev Immunol 13(4):227–242. doi:10.1038/nri3405

51. Qin H, Lerman B, Sakamaki I, Wei G, Cha SC, Rao SS, Qian J, Hailemichael Y, Nurieva R, Dwyer KC, Roth J, Yi Q, Overwijk WW, Kwak LW (2014) Generation of a new therapeutic peptide that depletes myeloid-derived suppressor cells in tumor-bearing mice. Nat Med 20(6):676–681

52. Srivastava MK, Zhu L, Harris-White M, Kar UK, Huang M, Johnson MF, Lee JM, Elashoff D, Strieter R, Dubinett S, Sharma S (2012) Myeloid suppressor cell depletion augments antitumor activity in lung cancer. PLoS ONE 7(7):e40677. doi:10.1371/journal.pone.0040677

53. Steding CE, Wu ST, Zhang Y, Jeng MH, Elzey BD, Kao C (2011) The role of interleukin-12 on modulating myeloid-derived suppressor cells, increasing overall survival and reducing metastasis. Immunology 133(2):221–238. doi:10.1111/j.1365-2567.2011.03429.x

54. Kerkar SP, Goldszmid RS, Muranski P, Chinnasamy D, Yu Z, Reger RN, Leonardi AJ, Morgan RA, Wang E, Marincola FM, Trinchieri G, Rosenberg SA, Restifo NP (2011) IL-12 triggers a programmatic change in dysfunctional myeloid-derived cells within mouse tumors. J Clin Invest 121(12):4746–4757

55. Kerkar SP, Leonardi AJ, van Panhuys N, Zhang L, Yu Z, Crompton JG, Pan JH, Palmer DC, Morgan RA, Rosenberg SA, Restifo NP (2013) Collapse of the tumor stroma is triggered by IL-12 induction of Fas. Mol Ther 21(7):1369–1377. doi:10.1038/mt.2013.58 mt201358 [pii]

Printed in the United States
By Bookmasters